PRAISE FOR
FUTURE PERFECT

"Sure it's premature to grant an award for all the 1980s, but I'm ready to take the plunge with **Future Perfect** by Stan Davis. His description of the firms and markets of the future has no close peers. Learn about the commanding role of time in management thinking, the growing importance of intangibles (e.g., information) in every product, and life in organizations without hierarchy."

Tom Peters

"Davis looks at elements of space and time in ways that help us better meet the future. . . . A brilliant work that should be read, and pondered, and pondered again."

The Stock Market Magazine

"Reversing what has become a common problem—late adoption of new technologies—is the central technological problem facing American industry. **Future Perfect** is a perfect place to begin that task."

Lester Thurow, Economics Dean
Sloan School of Management, MIT
and author of *The Zero-Sum Society*

"**Future Perfect** is easy to understand and packed with lots of powerful ideas. I enjoyed it immensely."

John Sculley
Chairman and CEO
Apple Computer, Inc.

"Perhaps a good definition of brilliance is the ability to look at the ordinary from new and different perspectives. That's what Davis does in this remarkable work."

The Stock Market Magazine

"As a writer and consultant, Stan Davis is one of the most skilled advisors today on what it takes to manage a corporation in the economy of tomorrow."

Walter V. Shipley, Chairman and CEO
Chemical Bank

"[Davis] demonstrates with a remarkable range of examples—past, present, and future—from a new perspective. Davis's future is provocative enough to cause even skeptics at least to consider what he has to say."

Booklist

"Offers corporate executives thoughtful, challenging perspectives that—for a refreshing change—stress process rather than prophecy."

Kirkus Reviews

"Managers will find this book valuable. . . . [Davis] is particularly effective in demonstrating that, today, the limit to innovation in business is not technology, but managers' obsolete viewpoints."

The Toronto Globe & Mail

Future

ADDISON-WESLEY PUBLISHING COMPANY, INC.

Perfect

Stanley M. Davis

Reading, Massachusetts   Menlo Park, California   New York

Don Mills, Ontario   Wokingham, England   Amsterdam   Bonn

Sydney   Singapore   Tokyo   Madrid   San Juan

Material appearing on pages 24–26 and 31–39 has been adapted, with only minor changes, from the author's article "Transforming Organizations: The Key to Strategy is Context." Reprinted, by permission of the publisher, from *Organizational Dynamics*, Winter 1982, © 1982 American Management Association, New York, all rights reserved.

Library of Congress Cataloging-in-Publication Data

Davis, Stanley M.
  Future perfect.

  Includes index.
  1. Management.   2. Organization. I. Title.
HD31.D325   1987   658   87-18669
ISBN 0-201-11513-1
ISBN 0-201-51793-0 (pbk.)

Cover design by Marge Anderson
Text design by Douglass Scott/WGBH Design
Illustrations by Barbara Davis
Set in 11-point Electra by DEKR Corp., Woburn, MA

ABCDEFGHIJ-DO-89
Second printing, February 1989
First paperback printing, August 1989

To my wife, Barbara

Contents

Acknowledgments

As with most any book, there are always many people to acknowledge for their contributions and support in making the work possible. In this instance, the greatest appreciation, by far, is due to an organization. The MAC Group is an international management consulting firm, with offices in ten cities around the world. MAC has been my most constant and enthusiastic supporter throughout many years. Larry Bennigson and Tom Howe, particularly, have believed in my vision of new management models for a new economy, and have encouraged the writing of this book with ideas, finances, staff, and project briefings. With the completion of this book, The MAC Group is moving the ideas into development activities that are deliverable as services to clients. Many other MAC professionals and staff have generously assisted me with ideas, references, feedback, and editorial work. Although too numerous to name all, particularly, I would like to thank Howard Schwartz, and also Bill Beizer, Janice Brodman, Anne Evans, Ed Goldstein, John Hall, Pierre Loewe, Michael Packer, Lucy Reid, Ed Rucker, and Sid Seamans.

Next, I owe a great deal of thanks to three people who taught me aspects of bringing out a book that were absolutely essential: Raphael Sagalyn, my literary agent;

William Patrick, my editor; and George Gibson, my marketing maven.

The Environmental Scanning Association generously provided some research funds, and Fred Foulkes and the Human Resource Policy Institute at Boston University, where I am Research Professor, have been very supportive. Anthony Athos and Professor Larry Greiner of the University of Southern California have been valuable advisors.

My ideas for *Future Perfect* began around 1977, during my work on Citicorp's ten-year strategic plan, where I had the pleasure of working with two very creative people, George Vojta, now EVP Bankers Trust, and Jacqueline Brandwynne-Cotsen now of Antebi & Cotsen in Los Angeles. Carlo Brumat of CEDEP in Fontainebleau, France, and Gianni Lorenzoni of University of Bologna, Italy, were also helpful in thinking through some of the basic conceptualizations. Robert Abramson of Primerica, John Humphrey of Forum Corporation, Howard Pifer of Putnam, Hayes and Bartlett, and Fred Smith of Federal Express also contributed in significant ways.

Many human resource professionals have been commentators and friends around this work; most particularly, Nelson Heyer, recently retired from IBM. I also am grateful to Jerry Dols of Chase Manhattan Bank and Arnold Kanarick of Analog Devices.

The quality of all this advice and counsel should not, of course, be put in doubt by any inaccurate meanderings that remain in theory or in fact, for which I take full responsibility.

Stanley M. Davis
Chestnut Hill, Massachusetts USA
September, 1987

Future Perfect

Aftermath

"Is there any other point to which you wish to draw my attention?"

"To the curious incident of the dog in the nighttime."

"The dog did nothing in the nighttime."

"That was the curious incident," remarked Sherlock Holmes.

—*Arthur Conan Doyle*

When I was young, I was terrified by science and bored by technology. I took only the required minimum of science courses to earn my liberal arts degree. Then, around ten years ago, in early middle-age, a strange thing happened. I found myself reading articles about science and technology. I was now gravitating toward what I had studiously avoided. I didn't know what was going on. Fortunately, I had enough sense to tell myself, "It doesn't yet matter why you're enjoying these readings, just keep on doing it and the answer will become clear."

And it did become clear. My field is management and organization, particularly organization structure at that time. I loved my work, yet I had become terribly bored. The subject seemed to say more and more about less and less. I would ask colleagues, "What are the most important questions that need to be asked and answered about organization structure?" The responses, from bright and interesting people, were narrow and dull. We were studying dusty nooks and crannies, not creating a structural revolution. By contrast, the physicists and biologists I was reading were asking very fundamental and exciting questions about structure.

I looked up "structure" in the dictionary, and it said "the interrelationships of the parts of a whole." This was a definition that applied equally to physics and to management – to atoms, to organizations, and to the universe. Physicists were talking about structure as packets of energy with the potential to interact. I wondered, how would we structure organizations if we looked at them as being made up of packets of energy (people, technology, and other resources) with the potential to interact? Surely, we would need a language different from static boxes and lines on organization charts.

At about the same time, I worked on Citicorp's ten-year strategic plan. In the United States, unlike Japan and parts of Europe, ten years represents very long range thinking, and so I had the opportunity to study causes and consequences of major economic transformations, past, present, and into the future.

These two events, unintended reading in science and long-range consulting in business, created the beginning of the inquiry that resulted in this book. I never imagined that I would write about science, technology, business, organization, and philosophy in a new economy. What a surprise I had! The next step was to see how new frameworks in science and technology illuminate what will become new models for business and organization.

We are bombarded every day with new technological developments, some so advanced that they challenge our ability to comprehend them. Computers that make logical inferences and process information the way our brains do. Others that take verbal dictation for up to 20,000 different words. Machines to walk up and down our chromosome ladders and read every gene. Products that are only a few atoms thick, others that are created in 72-dimensional space. There are also momentous advances producing supercheap everyday materials that will replace magnetic fields with superconducting fields and carry electricity without the slightest loss of energy. The marketplace has not caught up with these exciting new technologies, and organizations have an even longer way to go.

Technology provides a conceptual bridge that links the operating rules of science and the universe with those of economics and that ultimately gives shape to the way we manage. It is no accident that Sir Isaac Newton came before Adam Smith, whose theories in turn had to be spelled out before Henry Ford could create the assembly line and Alfred Sloan could then devise the divisional corporate structure. A basic progression governs the evolution of management in all market economies: fundamental properties of the *universe* are transformed into *scientific* understanding, then developed in new *technologies*, which are applied to create products and services for *business*, which then ultimately define our models of *organization*.

UNIVERSE ⟶ SCIENCE ⟶ TECHNOLOGY ⟶ BUSINESS ⟶ ORGANIZATION

The dilemma for managers is that dominant organization models are the last link in the progression to develop, and are not likely to occur until the shape of the economy is fairly mature. While the new economy is in the early decades of its unfolding, businesses continue to use organization models that were more appropriate to previous times than to current needs.

It wasn't until the 1920s that Sloan developed at General Motors what was to become the basic model for industrial organization – a decentralized operating system combined with centralized policy and financial control. This was more than one hundred and sixty years after the industrial revolution began in England, and sixty years after it had transformed America from an agrarian to an industrial economy. It was not introduced, in other words, until the twilight of the epoch, just twenty years before the industrial period drew to a close.

Now, thirty or forty years into a new, postindustrial economy, we find that Sloan's industrial model is still the major one used for organizing corporate America. We have industrially modeled organizations running postindustrial businesses. It is no wonder that we manage our way to economic decline. Our managerial models, the "context" in which we manage, don't suit the "content" of today's business.

· · ·

· · ·

· · ·

These nine dots are part of a familiar exercise in creating new contexts. The object is to connect all nine dots by drawing no more than four straight lines, without lifting the pencil from the paper. The key to the puzzle lies in your ability to redefine the context in which you see the nine dots – comparable to Sherlock Holmes hearing dogs that *don't* bark. There are many solutions, and the most common one is shown at the top of page 8.

Here is another: if the paper is folded along a line equidistant between the first two rows of dots, and also folded along another line equidistant between the second and third rows of dots, then the dots will touch each other in three triplets and connect along a single line. A third solution is to use a very big brush to cover all the dots with one broad stroke. A fourth solution comes from non-Euclidian geometry, in which parallel lines meet at infinity; therefore extend a line through one row to infinity, then zigzag back through the middle row, and back again through the last row. A fifth is that when you hold the page so that only its edge shows, all the dots are on that one same line.

After stumping thousands of executives in speeches and training programs, I showed the problem to my seventeen-year-old son Rick who offhandedly said, "Well, instead of moving the lines through the dots, why not move the paper, and pass the dots through one long line?"

The point is this: if we accept the boundary implied by the nine dots, we will never solve the problem. We must create a new context and, by that means, find new solutions.

What begin as insights often develop into models, which then evolve into techniques, after which they become deliverables. Through time, the deliverables get consigned to becoming conventional wisdom and artifacts, only to end up finally as inadequate. Ultimately, we initiate again the search for new insights. The insights that began the industrial revolution have run just this gamut, and now we need new models to manage the corporations in our new economy.

These new models first get articulated in our scientific and technological understanding of how the universe works. My intention in this book is to give new meaning to time, space, and matter in shaping tomorrow's business and organization. In the industrial economy managers considered time, space, and matter as *constraints*, whereas in the new economy they will come to think of them as *resources*. This will require profound transformations in the way we think about time, space, and matter. Just as the scientific shift from the mechanistic age of Newton to the holistic age of Einstein affected notions of what was meant by time, space, and matter, these new notions in turn will affect the managerial transformation from an industrial mindset to a fundamentally new one.

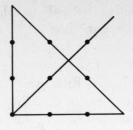

Young people beginning careers in the 1980s will be middle managers as we enter the new millenium, and the major force in business during the decades after that. They will urgently need the frameworks that are appropriate to this future reality just as CEOs do today. We, who are in our middle and later years must not just pass on to them the exhausted models of the last economy. We owe them more than that.

We critically need new management theory to explain and to further the transformations that we are witnessing yet only dimly understand because we are too much in their midst. In the industrial economy, our models helped us to manage aftermath, the consequences of events that had already happened. In this new economy, however, we must learn to manage the beforemath; that is, the consequences of events that have not yet occurred. This is managing in the *future perfect* tense. By 2001, when the new economy probably will have matured, we will observe our holistic approach to management and wonder how it ever could have been otherwise.

Any Time

"The trouble with our times
is that the future
is not what it used to be."

—*Paul Valery*

The shortest provable unit of time is 1^{-43} second (that is, .0000000000000000000000000000000000000000001 of a second), the life span of the briefest elementary particle, the smallest known sub-atomic matter in existence. At the other end of the spectrum, the longest known unit of time is the nearly twenty billion years that have elapsed since the Big Bang created the universe. The solar system is about five billion years old, the earliest life on earth is around three billion years old, and the specie man has existed for approximately two million years.

In business, time is measured on a very narrow portion of the total spectrum. Few major corporations can be counted by centuries, careers are measured by decades, and products by years. Accounts are generally payable monthly, people often work nine to five, coffee breaks are fifteen minutes, push-button phones save you seven seconds dialing time, and lasers work in nanoseconds (one billionth of a second).

In other words, time is a way to measure, and hence define, existence. And these definitions are culturally imposed. The calendar is an arbitrary system for reckoning the beginning, length, and divisions of a year. There are the Chinese, Gregorian, Hebrew, Hindu, Julian, Revolutionary, Roman, and many other such calendars. The Gregorian calendar, for example, now used in most parts of the world, was adopted for use here as recently as 1752.

Isolation experiments in deep caves have shown that man's natural periodicity is closer to 24.5 hours. If man did not adjust his natural circadian rhythm to his culturally determined yearly calendar, then daily life would get messed up pretty quickly. In other words, the things that we choose to measure and the way that we measure them tell us lots about what we value and how we see the world. This is true for business executives as well as for astronomers.

In the early 1900s, roughly half way through the industrial era, U.S. efficiency expert Frederick W. Taylor invented time management. In the industrial model of time, a worker doing endless drudgery, where one day is just like the next and nothing will ever change, takes the point of view that time is cyclical, an endless repetition of events. The clockwatcher who celebrates TGIF, and the harassed manager who only has two hours to get the report out, see themselves as stationary "in" time, with the future moving toward them rather than vice versa. Alvin Toffler humorously expresses this perspective: "Future shock is the dizzying disorientation brought on by the premature arrival of the future." The managers who prepare for next week's conference, who launch next season's product line, and who chart their career development take the approach that time is a one-way street that we move along; we advance toward the future, not it toward us.

The business model of time in the industrial world was very much from the corporation's perspective. The focus was on internal reorientations and actions. Even "9-to-5 time," defining as it does a regularly recurring event, is from the industrial perspective of the *producer*, rather than the postindustrial perspective of the *consumer*. When the corporation focuses externally, on the customer, it transforms its sense of time.

A new economy manager once said to me, "Customers use our time up until their decision to buy, after that we are using *their* time. Therefore, we must deliver immediately." The key, then, is the shortening of the elapsed interval between the customer's identified need and his, her, or its fulfillment.

Over five years, Matsushita's refrigeration factory reduced its average manufacture time from 360 to less than three hours. In the late 1970s, Toyota found that more costs and more time were expended by a car in the distribution system than in the factory. The manufacturing company absorbed the sales company, and Toyota today is working to deliver cars to customers within a day of receiving a dealer's order.

At Stratus Corporation, when their customer's mainframe computer malfunctions and the back-up part takes over, the factory automatically receives a signal from the computer that it needs a new back-up part. The first time the customer knows anything is awry is when the replacement part arrives by Federal Express the next morning.

There are thousands of services, such as making a pair of eyeglasses or developing a roll of film, whose customer waiting time can and has been cut from days to minutes. Every business can chart the elapsed time for every step *from conception to consumption* and work to reduce it. Marginal reductions (10–20 percent) generally can be accomplished by improving efficiencies; reductions in multiples (50–100+ percent) generally require reconceptualizing the production, distribution, and/or delivery processes themselves.

The transformative quality lies in the elimination of

any waiting at all – "zero-based time." Whenever the customer needs the product or service offered, it should be immediately available. Not during working hours the next day, or after the weekend, or as soon as the next representative is free; but any time, instantaneously.

Our emphasis here is different from common-sense speed. We are not merely suggesting alacrity. We are talking about *instantaneous* products and services, those that are offered within the blink of an eye of their conception. If you can imagine this occurring, then the product is in research, in development, in manufacture, and being consumed *virtually* all at the same time. This is truly a holistic conception of the product.

The ability to do this seems to have more to do with technology than with tangibility. High tech businesses, whether they make products or offer services, operate more in a real-time mode than do low tech businesses. In Silicon Valley, for example, companies are toiling to bring out their next-generation work stations practically from the time they are shipping the prior-generation model. It's hard to imagine that with a toaster. Brokers have embraced information technology, and they can put together a product and be selling it within the week. Insurance companies have been slower to absorb the technology and virtually never generate new products this quickly.

Speaking practically, whatever your business, think about how you can create products and services in real-time that you can deliver instantly. Even in the slowest moving company, this contextual shift will speed things up.

In most hotels, for example, check-out time is about 1:00 P.M., and check-in time is about 3:00 P.M. The logic is that the hotel needs the two hours to clean and prepare the rooms for the next guests. Obviously, however, they do not clean all of the rooms within that two-hour stretch; they begin as soon as guests start leaving in the morning, and finish later in the afternoon. In other words, room preparation is going on most of the day. The time between check-out and check-in is the lag-time the hotel needs because it thinks it cannot operate in real-time.

Imagine, then, a hotel chain that cuts the two hours to one hour – while it is getting its systems in order – and then a few months later cuts the lag-time to zero. This requires systems and schedules to clean rooms in real-time. Its advertising campaign announces, "Check-in any time, check-out any time. Unlike the competition, we operate on your schedule, not on ours!" This is no different from what car rental companies do all the time. Guests are charged for the minimum equivalent of a 24-hour stay, and pricing is according to usage. Technically, there is little to prevent hotels from making this switch. What it requires is a shift in the context: to view time as a resource in the new economy, not a constraint.

The larger rules operating here are:

● Consumers need products and services ANY TIME (i.e., in their time frame, not the providers').
● Producers who deliver their products and services in REAL-TIME, relative to their competitors, will have a decided advantage.
● Operating in real-time means no LAG-TIME between identification and fulfillment of the need.

In banking, too, the customer gains by being able to bank "any time." Less than a century ago "bankers' hours" generally meant from 9 to 2. Gradually banks extended their hours and now, with the advent of automatic teller machines (ATMs) in the 1980s, most banking can be done literally any time of the day or night. No longer need we stand in line to cash a check, waiting for a clerk to take down the information from two forms of identification. Today, our credit card can be inserted into a little credit-checking machine by the register, and MasterCard's Banknet internal telecommunications system will clear such a transaction in only two seconds in the U.S. and eleven seconds worldwide.

Real-time banking can work in the bank's favor, though, in elimination of the float. Float is the period of time, to take one instance, from when a customer writes a check until the check is cashed, cleared, and deducted from the account it was written on. Traditionally, that period was at least several days, and often over a week. During the time that the money is "floating" it is still accumulating interest for the check writer. The faster the transaction clears, the quicker the customer loses the float. This adds up to millions of dollars in revenues on billions of dollars of float at any given moment.

Thus, in the economy of any time, any place, no-matter, the debit card may begin to replace the credit card. At point of sale banking, consumers' bank accounts are debited immediately as they ring up their vegetables at the supermarket cash register. The same instantaneous approach to time is occurring on the corporate side. According to one major bank, payment terms of 2 percent per 10 minutes is becoming standard as electronic check-collection and payments dribble on-stream.

In an effort to be more competitive, banks are teaching their clients to make instantaneous electronic banking work for them. Citicorp is by far the most electronically advanced financial institution, and its "Citi of Tomorrow" and "Global Electronic Banking" advertisements emphasize the new meaning given to the old notion that "time is money." Here, for example, is one of its typical ads:

An Orient Overseas ship must pay a fee before it enters the Suez Canal. If payment begins when the ship arrives, the ship waits at the Canal — at a daily cost of almost $16,000 — until funds pass through correspondent banks. The other choice: Begin payments 8 to 9 days before arrival and lose the float.

Solution: Orient Overseas electronically instructs Citibank New York to transfer funds to Citibank Port Said. Citibank then provides same-day value funds and the ship proceeds.

To learn more, call an account manager.

Arbitrage is another financial activity that would not be possible unless the arbs could operate in real-time. The goal in arbitrage is to take advantage of price differentials associated with identical assets. This is done by tracking and trading simultaneously in stock futures and in a stock index that contains the company whose stock future has been targeted. Currently, this involves the sophisticated tracking of a half-dozen markets by means of multiple computers and monitoring screens, and simultaneously trading overvalued futures against the current undervalued underlying index. And when stock quotes are available on a real-time basis, as is already the case with futures and commodities markets quotes, then, theoretically, even small investors can do arbitrage. Arbitrage cannot be done unless time is as critical a resource to manage as the information itself.

In the industrial economic model, the adage "time is money" implied that money is the key resource and should not be squandered, and that time is the way to measure whether or not the resource is being used wisely. In the new economic model, we could reverse the statement to read, "money is time." Time is the key resource, and money the way to measure whether or not you are getting as much full value out of it as you might. The resource and its measurement are reversed. Walter Wriston, retired chairman of Citicorp, was thinking this way when he said that banking was no longer about money, but about the "time value of information."

To move from conception to consumption instantaneously is a hypothetical extreme. But one gains competitive advantage simply by being able to move from identifying to satisfying market needs faster than one's competitors – faster than the "industry" leader and/or the "industry" average. Not all moves to instantaneity will be equally valued by all market segments, and time should join the ranks of price, quality, and service in determining market niches.

Just how fast you can move depends on what business you are in, what function within that business you are talking about, what economy that business operates within, and also what unit of time you are measuring. Managers should assess the time occupied at each point in the value-added chain (e.g., research $\longrightarrow$ development $\longrightarrow$ manufacture $\longrightarrow$ distribution $\longrightarrow$ sales) and focus attention on shortening the time it takes to perform those functions at the point where the value-added is the greatest in their particular business. There is competitive advantage in providing the same product or service, at the same price, in 20 percent less time.

The relevant unit of time – minutes or months – varies with each business. How quickly, for example, can a product be brought from the drawing board to the market? How long does it take to assemble the car, garment, or house? In the industrial economy, for example, inspection on the assembly line was largely performed on a random sampling basis, because of cost and lost time. In the new economy, machine vision permits inspection of every part in real-time, as it is being produced – and the machines don't tire, lose concentration, get bored, or apply subjective standards.

How long after an accident occurs is the claim settled? How fast can the bank provide an answer on financing a mortgage? Does the customer have to wait five days? ten days? twenty days? And if the bank fails to deliver on time, will it pay for the appraisal? How long does the customer have to wait in line? It is particularly irritating for people who work that service companies will guarantee a date but are unable to specify a particular time for pickup or delivery of what needs service or repair. Most people cannot take the day off to wait around for a representative of a service company that cannot operate in real-time.

The relevant unit of time also varies with each type of economy. In agrarian economies it took a relatively long time to produce an item, and shortening that time did not significantly reduce the cost. The manufacturing sector therefore produced customized goods in low volumes. Production time was long, relative to that of an industrial economy.

By contrast, the technologies that led to an industrial economy created a manufacturing system based on high volumes and long production runs of standardized goods. Lower-cost goods were available to more people precisely because the price per unit was brought down by cookie-cutter manufacturing. It took a long time, and was very expensive, to tool and set up machinery for a production run, and you therefore had long runs with little change. Industrial production changes jeopardize capacity and reduce efficiency. The annual model change was a radical innovation in the automobile industry in the 1920s.

The relevant unit of time is also a matter of what elapsed interval you are measuring. Today, the annual model change is a less relevant index of innovation than is how long it takes to move from creation to consumption. Changing an ashtray on a Detroit-built car would take a few years of planning and execution before the new car owner got the different model. In the securities industry, by contrast, new products may move from idea to sale in a matter of weeks.

The ultimate logic is to shorten the elapsed interval to zero. In the no lag-time world, *ideas are acts*. One office-of-the-future company is working on this kind of time-space shortening, so that many machines no longer need physical contact to be given commands like "on" or "off." Some are sound activated; others are being developed to operate by eye movements. Apple marketed the mouse. Perhaps this new method will be called "mouse eyes." These technological developments are only a few years from widespread use.

Earlier futurist authors understood the negative power of such high tech worlds when they wrote novels about "thought police." On the positive side, however, when such innovations occur, new ideas become new businesses overnight. There is an increase in entrepreneurship. Products can be tested more quickly, and market response is measured instantly, as in the Nielsen ratings. For the knowledge worker of the future, ideas are actions.

The new technologies allow instant resetting of specifications, with virtually no machine downtime. CAD/CAM is the dynamic duo that reduces the costs of flexibility, and allows customization to defined market segments. In the garment business, for example, the greatest value-added lies in the fabric cutting. Telling industrial machinery to cut a 54 stout or a size 8 petite was not an economical proposition. Too few people wear these sizes to justify the machine adjustment time or cost. Laser cutting of garments and computerized controls, however, enable these sizes, as well as all others on the human spectrum, to be cut in small volumes and without the lengthy and hence costly delays. Specifications are reset instantly, with no machine downtime. This is a crucial element in what is being called flexible-system production.

Gannett Company applies this orientation – *from ideas to action in real-time* – in publishing *USA Today*. The relevant time gap in newspaper manufacture is the interval between the newsroom deadline and the first press run. The later these are, and the closer together they are, the more flexible a newspaper can be in deciding what to include in its morning edition. The industry average is Last-In at 11:00 P.M. and First-Out at 1:00 A.M. Gannett has managed to get their printer's version of LIFO down to 2:00 A.M. and 3:00 A.M., respectively, making them a clear industry leader.

Nothing demonstrates this advance so clearly as the fact that, in computer software, manufacturing disappears completely. The functions leap from design and development straight to consumption. Manufacture takes little more time than pressing the "return" button on the computer to make multiple copies for users. As any "hacker" or "pirate" knows, this kind of manufacturing can easily be done on one's personal computer.

Computer-aided transcription (CAT) offers another example of manufacturing in real-time in the new economy. Mechanical stenotype machines were introduced in U.S. courtrooms in 1910, with little change until 1986, when experimental CAT systems were introduced in Chicago, Detroit, and Phoenix. With these systems, an unedited transcript largely in readable English appears on IBM PC XT computer screens in front of the judge and both lawyers virtually as testimony is being given. The court stenographer presses a 22-button keyboard, writing phonetically. Keystrokes are fed into a computer that expands the shorthand into English, flashes it on the screens, and prints a transcript – all in real-time. In the U.S.'s litigious society, which can produce thousands of pages of transcribed testimony in a single case, this speeds up the system considerably.

In 1964, 80 percent of computer costs were in the hardware; and only 20 percent were in the software. In that world, manufacturing was a real constraint. Twenty years later, however, the proportions were reversed: hardware costs 20 percent; and software, 80 percent. Hardware manufacture is nominalized and robotized. Software costs quadrupled, but where are those costs located? The answer is basically in development costs, in the people and machines who are writing the new software. Once written, however, they hardly need to pass through a factory to get into consumers' hands. In other words, in contrast with software development and hardware manufacture, software manufacturing is a one-time thing, in other words, costs virtually never increase.

Real-time means that responses to inputs are fast enough to affect subsequent inputs and to guide the process. If A and B are linked together, then specific inputs to A should alter B almost simultaneously. In earlier technologies there were time lags between events A and B. Electronics introduced the real-time world of almost no time lags, and photonics will bring us there completely. But while today's technologies operate in real-time, today's organizations do not.

The unwanted time lags between inputs and outcomes are especially evident in the relationship between business and organization, particularly in the element of business that has to do with strategy. Strategy is a way of using time in a new context, as a resource. This allows you to see your business and organization in the future, interpolate your way backward into the present reality, and then manage your implementation more powerfully.

The adage, "You have to know what you want to do before you can do it," contains the definition of both strategy and organization – knowing what to do, and knowing how to do it. Strategy is the plan for future survival. Organization is the current arrangement for day-to-day application. In principle, the relation between the two is that a team, company, army, or nation should be organized in the manner that will best implement the strategy. A good strategy with poor organization is a thoroughbred without a rider, trainer, stable, or track. In principle, strategy precedes organization and the two are closely related; in practice, often they are not.

In the industrial context, *organization always lags behind strategy*. Because of the assumption that you have to know what it is you want to do before you can know how to do it, all organizations based on the industrial model are created for businesses that either no longer exist or are in the process of going out of existence!

That is a terrible state of affairs. The inherent weakness of this model, which developed in the industrial economy, is that no organization can ever be in sync with time, or totally appropriate for carrying out its mission or purpose. The mission, objectives, and strategy of the business in a mature industrial context will always come first; they will always be ahead of the organization. The time lag between formulating a plan and implementing it may be thought of as the distance between a strategy and its appropriate form of organization.

Organizations can do no better than catch up with the present, and there is even a Catch-22 to catching up: when you get there, "there" isn't there any more. Strategy is always focused on the future, but it is rooted in the present, or even in the past, if management is inefficient.

The name of the game, managerially and organizationally, is to catch up as quickly as possible. The shorter the time lag between strategy and organization, the more efficient is the business. Reduce the time lag by which organization follows strategy and, all else being equal, you will increase your success by whatever measurements you choose. This is a very inert conception of organization.

How can an organization implement its strategic plan with actions that are appropriate to the present-future rather than ones that are catching up with the past-present? The past, present, and future have to do, of course, with our traditional concept of time. A key element in implementing strategy in the context of the new economy involves a very different sense of time. The difference is akin to shifting from a Newtonian sense of time as absolute to an Einsteinian sense of time's relativity. Initially, this requires a rather abstract presentation of some fundamentals, which are necessary to state if we are to apply them in implementing strategic plans.

The sense of time that executives employ with the industrial model is to use the present organization as the vehicle for getting to the future, to the objective. At first glance this is logical. "What other organization can we use? It's the only one we've got." In the context of the new economy, leaders operate from a different sense – from a different place in time. Implementing a strategic plan in the new context will require operating with a different concept of time.

The only way an organization's leaders can get there (the objectives of the strategy) from here (the current organization) is *to lead from a place in time that assumes you are already there, and that is determined even though it hasn't happened yet*. This sounds bizarre or obscure only at first reading.

To explain this different sense of time requires a brief entry into the world of phenomenology and of semantics.

Here I am drawing on the work of George Herbert Mead, Gordon Allport, B. F. Skinner, A. Schutz, and Karl Weick. Weick holds that "an action can become an object of attention only *after* it has occurred. While it is occurring, it cannot be noticed." If everything is retrospective, then how are we to account for the fact that people and organizations plan and guide their actions according to their plans? Weick say that "even though a plan appears to be something oriented solely to the future, in fact it also has been accomplished. . . . The actor visualizes the completed act, not the component actions that will bring about the completion." This last sentence may be restated to read, "The manager visualizes the completed strategy before visualizing the component actions that will bring about the completion."

Schutz names this sense of time explicitly in Karl Weick's *The Social Psychology of Organizing*:

> . . . *The actor projects this action as if it were already over and done with and lying in the past. . . . Strangely enough, therefore, because it is pictured as completed, the planned act bears the temporal character of pastness. . . . The fact that it is thus pictured as if it were simultaneously past and future can be taken care of by saying that it is thought of in the future-perfect tense.* [1]

The implementation of a strategy has to be considered in the future-perfect tense. Using this time perspective, *the present is the past of the future*, and organization can be used to *push* the strategy toward its realization rather than be *pulled* along by it.

Organizations can implement strategic plans without having to be pulled along, and entrepreneurs do so regularly. Recently, I was working with a large client organization that had become very bureaucratic and slow to act. This organization had been started early in the century by one of America's giant entrepreneurs. While contrasting the founder's and current management's senses of strategy, a major difference became evident to me. The founder's strategy came out of his actions. Although today we can state the specific elements of the founder's strategy, he was nowhere so analytical. He was the proverbial man of action, a great visionary, and he built the biggest company in his business, in the world at that time. Subsequent management stood the relationship on its head, believing that their actions should come out of their strategy. The result was bureaucracy and, ultimately, one of the most major takeover actions in the corporate world.

I see the same "action before strategy" approach with most of today's entrepreneurs. They have a vision, often not articulated in anything so explicit as what we now call strategy. Their intuition is generally ahead of their conceptual framework, and they evolve a coherent and post-facto rationale for the details of what they are already doing. Managers see the causality flowing from strategy to action, while entrepreneurs see the flow in reverse.

For those whose strategy flows from their actions, rather than vice versa, strategy is the codification of what has already taken place; it is the writing of future history. This orientation places strategy in time past rather than in the future, where the formal industrial models would have it.

The lesson is that if strategy is the codification of what has already taken place, then it is the enemy of innovation. Organizations that foster innovation are not to be wedded to strategy as formal planning, but to strategy as intuition.

Intuitive strategy is nevertheless logical and conscious. It would be a romantic extrapolation to say that it occurs only in the creative right brain and not in the analytical left brain. From research done on chess grand masters, the fruit flies of artificial intelligence, we know that what appears to be intuition is really no more than a built-in recognition of habitual patterns, much like the one all of us build in language patterns. Managing in real-time, therefore, means building in strategic objectives sufficiently enough so that implementing actions seems to flow intuitively from their recognition.

Strategic planning is the one new business function in decades to have captured the serious attention of senior management. It is also the one business function that has made time intrinsic to its modeling. When strategic planning was in its infancy, in the early decades of the new economy, it was little more than extrapolation of the past into the future. As the field became more sophisticated, practitioners engaged in various assessments of that likely future, as in the environmental scans and future scenarios. From these tools a clearer statement of strategy emerged: "Here is what our company is going to look like in the future." Thus, interpolation replaced extrapolation in strategic planning: "If this is what our company looks like today, and that is what we intend it to look like 'X' years from now, then we can know *what* changes must be made to become that newly described entity."

Harold Geneen expresses a similar thought in his book *Managing*, "A three sentence course on Business Management: You read a book from the beginning to the end. You run a business the opposite way. You start with the end, and then you do everything you must to reach it."[2]

This is one way of describing what has since become the very sophisticated area of strategic planning and the formulation of strategy. The implementation of strategy, however, is still not that sophisticated, and we can make important advances by adding two additional orientations: *strategic control* and *organization strategy*.

Strategic control combines elements of both the strategic planning and the control functions. Most strategic plans take a one-time view of the future, "X" number of years out, and the better ones monitor the "X" on a rolling basis. Control systems monitor the past; what has already happened, and did you do what you said you would do? Strategic control takes the tracking and checking-up characteristics of the control function and, rather than locating them in what has already happened, it places them in the future. It continually tracks how the future "X" is changing as you get closer to it, so that, although you are still managing to stated future objectives, the objectives are updated daily to correspond to the shifting reality. This is much like the tracking system in a launched missile that corrects in flight for the evasive tactics of the target plane out there, ahead of it. When you hear that Merrill Lynch has bought and then sold a real estate business, for example, instead of interpreting this as dumping a bad investment, you may see it as an appropriate redeployment to changed futures. This is strategic control.

Organization strategy is another tool that should be added to management's kit. Remember the linear logic of the industrial model: organization lags behind strategy. Where strategic planning has become interpolative, organization planning, to the extent that it exists at all, is still largely extrapolative:

"*Our sales have grown at X percent for the past ten years, and the number of our employees has grown at 80 percent of X during that same period. Given our expected size at time . . ., we will therefore have to hire . . . people during the next . . . years.*"

Planning the future organization should be accomplished in the same way that the strategy for the future is determined – interpolatively. Each element of organization that will be appropriate for the future should be spelled out in detail. For example:

"Given the kind of business that we intend to be, what are the appropriate structures, systems, people, and corporate values for that business? How do they differ from the current ones? What steps are necessary to move them from here to there?"

A business strategy for the future needs its counterpart in an organization strategy for that future business. The not-yet-existing organization is no more nor less real than the not-yet-existing business which the strategy envisions.

All this sounds great in theory, so why doesn't it seem to happen in real life? Because people have a vested interest in continuing to see time as a restraint, rather than as a resource. By doing so, they have created a role for themselves. People who identify problems generally identify themselves as problem solvers, yet the irony is that they then have a stake in the problem staying identified but unsolved. They adopt the posture that the problem is so large the best they can do is whittle away at it.

Socially meaningful lives can be devoted to curbing addiction, bureaucracy, crime, disease, and so on through the alphabet. What all these people share in common is a baseline presumption -- the problems are so great that totally eliminating them is an absurd and ridiculous thought.

If you are beginning to penetrate, you are treated as

dangerous and subversive. And it is subversive. It subverts the context on which the problem solver has built a career, on which the professionals have built their organizations, and on which the society has built its institutions. If the problem is actually eliminated, totally, then the need for the services of the problem solver is also eliminated. How low does the crime rate have to get before it is a threat to law enforcement agencies? How many people have to get jobs and come off the welfare rolls before they are a threat to social work agencies?

Professions as varied as medical doctors and personnel managers whittle away at problems that will never disappear, in large part because they do not function with a real-time orientation.

The industrial context of modern medicine, for example, is the cure of sickness. The overwhelming majority of time, money, and personnel is therefore devoted to remedial medicine rather than preventive medicine. This is why doctors know so much about illness and so little about health. In fact, the only way we have to define health is in terms of the absence of illness. That is comparable to defining the economy only in terms of debt, never balance or surplus, and business only in terms of loss, never profits or redistribution. Anyone who can define health can transform the context of medicine. Doing so requires seeing time – real-time – as a resource.

Personnel officers in corporations also often share the same problem. The context in which personnel managers usually hold their jobs was aptly expressed to me by one such personnel figure, a senior vice-president in a Fortune 500 company. He said,

> *Personnel people can be the conscience of the corporation; they can say, "the King has no clothes"; they*

*can call bluffs, speak truths, speak out for good
against evil, and care for people. They can do all of
this with support and impunity because, even though
they don't acknowledge it to themselves, they have
made a psychological contract with the corporation
that they will never win.*

What would the personnel function look like if those
who occupy the role transformed the context? How
would it affect people's behavior? One function for
which a personnel department is responsible is the organization's reward system. Reward systems in almost all
organizations are built on the context of "what is not" –
if you do better you will get more. Rewards in such a
system are accumulated, as in Hay points; they are
earned, as in salaries; and they are taken away, as in
cutbacks and demotions. As one management psychol-
ogist once put it, most are based on the carrot-and-stick
mentality, and what stands between the carrot and the
stick is the jackass that the system is designed for.

What would a reward system built on "what is" look
like? It would begin in the future, with the mission,
objectives, and strategy of the corporation, related to
each business unit. Every organization member – not
just the leadership – would be very clear about how his
or her job implements that future.

Currently, how many employees carry out their jobs
every day with a clear idea of the organization's fundamental purpose? How much stronger would an organization be if they had this perception, and what would
the organization look like? One thing is certain, it would
reduce an emphasis on hierarchy. Each job is of equal
importance, both to the organization and to all individuals in it. Put another way, the most important job is
the one that is not being done. Doing the job is the
reward in such a system. The more this is so, the more

powerful is that organization. That is a system whose reward is wisdom. Few reward systems are premised on this real-time context.

When time becomes viewed as an intrinsic dimension of something, it gets treated more as a resource to be drawn upon and less as a constraint to be gotten around. Einstein added time to scientific models of the universe; then electronics used time resourcefully in technological models; strategy added it next to business models; and now we should make time an intrinsic dimension of our organization models. Until this occurs, until organizations adapt to change in real-time, is organizational lag necessary?

Theoretically, there must be some lag. Practically, there should be as little lag as possible. Remember, strategy tells you what the business is going to look like, organization tells you how you are going to get it to be that way. Once the organization is "that" way, it has implemented the strategy; and when this has occurred, "that" is no longer the strategy.

By definition, there is no organization whose culture, structure, systems, and people are completely appropriate for its strategy. If all these components of the organization were completely appropriate, the strategy would be realized; that is to say, it would be operational and no longer strategic. Successful strategy self-destructs. An objective, once accomplished, is no longer an objective. The realization of strategy is always futuristic. Because organization is the mechanism for implementing strategy, for realizing the future, time must be an intrinsic dimension to it.

When time is extrinsic it becomes a constraint to realizing the organization's objectives. Treated this way, organizations exist only in the current, inadequate, and unresponsive framework. Literature on organization, organization consultants, and most personnel, for example, deal with organization from a remedial point of view: how to cure what is wrong with it, how to make it better, how to get it to somewhere that now it is not. The organization is thought of as some lethargic giant that never quite does what you want it to do, and you have to pull and tug at it to get it to go where you want it to be. Those who focus on organization in each case have a sense of what it should be, and when they look at the current organization, their conclusion is, "This is not it."

From this viewpoint, the organization retards the implementation of strategy. The valence is always negative; it is only a question of how much. Reduce the negatives, remove the impediments, improve the organization, and the best you will do is reduce the lag between the formulation of strategy and its realization. It is not possible to create real-time organizations while treating time for implementation as a constraint. Most restructurings, management development programs, reward systems, information processing techniques, and other elements of organization are aimed at such improvements.

Any improvement is premised on the supposition that what exists is what is not wanted. Even though individual executives may not lean toward philosophy, this is nevertheless a philosophical anchorage point from which most tend to proceed. The preindustrial world view tended to accept "God's will" and a natural order. The view of the world that developed and nurtured the industrial economy, however, sought to predict and control nature. Things could be improved. Millions of dollars are therefore spent improving the organization to move it in the direction of the announced strategy. And success is always measured by degree – more or less – but never complete.

Our industrial frameworks come from a mechanistic world view that is forever tampering with and improving on what is. We usually view the present organization as wrong, and the right one exists only in the past or in the future. Maybe the present organization is only a little bit wrong; maybe only a very, very little bit wrong; but never, "This is it." All effort, whether it be good works in religion or outstanding achievement in business, is an overlay on the fundamental belief that, currently, "This is not it."

This is comparable to the distinction between invention and discovery; unlike invention, the object of the discovery exists before anyone knows it. People discover their humanity, they don't invent it. Would you, for example, rather work for a boss who discovers you or invents you?

Michelangelo's approach to sculpture was similar to the act of discovery, whereas the ordinary artist invents, by carving a figure into the stone. In other words, the ordinary artist, believing that the figure did not exist before he created it, approaches the stone from the context of inventing the figure that is not there. Michelangelo, however, began with the assumption that the figure was in the stone before he touched it. His job was to uncover the figure that was already there. His statues of the slaves, alongside his David in Florence, are the best examples of this approach. Only part of the figures are visible; the rest are enslaved by the stone. His genius is that anyone looking at the statues knows that the rest of each body is there within the encumbering rock.

The effective organization, particularly its leadership, understands that it has already succeeded ("This is it."). The only problem is that not everybody in the current organization knows this. If we assume, for example, that the new organization is out there to be gotten, then we don't have it, and in fact we never will. If we start from the context, "The way we behave is inappropriate; instead we must learn how to behave like . . . ," then the message we are putting out is that we are not implementing the strategy.

If, on the other hand, we start from the context, "The way we behave is appropriate to the strategy," then the membership in the organization will know that their goals are being accomplished each moment, in real-time. Each meeting, each decision, each activity is confirmation that the new organization "is." It already exists.

Executives who lead from an orientation that what they want for the organization lies "out there" can be only as powerful as the never-realized future. Time is their constraint, not their resource; they are less powerful than they might be. By contrast, those who lead an organization from this context are powerful because they already have what they want. Notable examples come from major political, religious, and business leaders. Lenin and Mao knew that the revolution had already taken place in Russia and China, respectively. Revolution is time transformed.

Long before these leaders reached the pinnacle of organizational power, while they were still considered fugitives by those who ruled, they knew they had succeeded. All that remained, however great the task and whatever the cost, was to execute whatever steps were necessary for others to accept the new reality. From the position that the revolution has already occurred, and that a return to the earlier time is impossible, hardships are taken as tests and signs that strengthen rather than diminish the new order.

In only a few years it will probably become fashionable for corporations to issue long-range plans for the year 2000, and these kinds of plans should address the new economic context we have been discussing. Managers will have implemented a long-range plan when every action taken is both discovery and implementation of the content. It is management as source, not as outcome. Managing this way takes place with a mental orientation in the future perfect tense.

These different orientations toward time can be understood better if we differentiate between two points (see Figure 1):

1. The point that separates the time before and after the decision has been *made* and the resources have been allocated.
2. The point that separates the time before and after the decision has been *implemented*.

Figure 1

Time Line Context

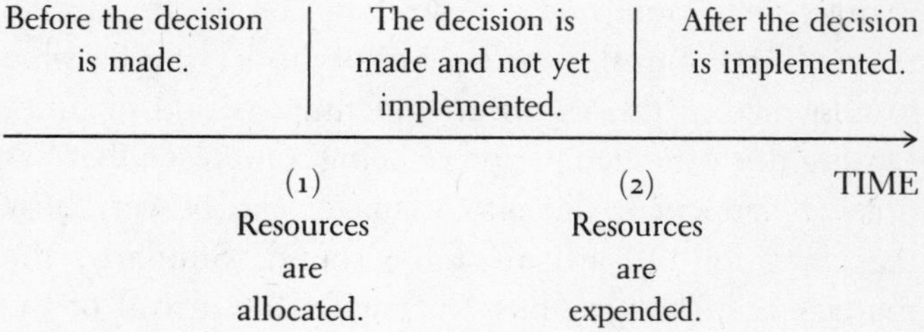

Before the decision is made.	The decision is made and not yet implemented.	After the decision is implemented.
	(1) Resources are allocated.	(2) Resources are expended. TIME

It is between these two points in time that management seems to have the greatest difficulty: between (1) an effective decision and (2) its efficient implementation. The context in which most managers hold this middle span in time makes it very difficult for them to get from (1) to (2).

Operating managers generally begin in the present with the already-made decision (1) and work toward the future (2). Managing from a different context, they would mentally operate from point (2), working to realize what they know is already so, even though it hasn't happened yet.

To understand this transformed notion of time, let's take examples from the physical world: sound and light traveling through time. Sound travels at the speed of about 660 miles per hour. Think of two airplanes, one subsonic and the other supersonic, traveling from point (1) to (2). The sound emitted from the subsonic plane reaches point (2) at the same time the plane does. The supersonic plane, however, reaches point (2) before its sound does.

Imagine arriving in a plane at (2) and then waiting for the arrival of your own sound. In this context, you are there before the fact. You have created a phenomenon, gotten ahead of it after it was created, and observed it catch up with you.

In this example of sound traveling through time, the lag between the two points in time is so brief that there is not much more one can do than observe the occurrence. But if one takes the essence of the act, it is possible to conceive of the lag as an indefinite period of time. During this extended period of being someplace that has not yet happened, the pilot/manager can be very busy preparing for the arrival of the sound. Similarly, the manager can be very busy managing the arrival of the organization that will be appropriate for the time when it arrives.

Light traveling through time operates by the same principles, even though our experience of it is different because of its speed. Light travels at 186,000 miles per second. Einstein's principle of relativity enables us to comprehend what it means for light to get ahead of itself. We know that the starlight we see is the light of stars that no longer exist. It is leftover light, aftermath.

If aftermath is the effect of past action, as I said in the beginning of the book, we can speak of beforemath as the effect of actions that have not yet occurred. Anxiety about a forthcoming event is a familiar example of mismanaging beforemath. Implementing a strategy involves managing in the future perfect tense – that is, managing the beforemath. People who take out life insurance and have home mortgages are managing the beforemath – they are managing the consequences of events that have not yet taken place.

Surely if science can renew a dialogue with religion that was unattended for centuries, then management can derive benefit from the farthest horizons of science as much as from science's more familiar territories.

We need not go galaxies away, however, to find examples of these different perspectives for managing time. People considering several alternative jobs in pursuit of successful careers have not yet committed their resources. Their options are open. At Point (1) in Figure 1, they choose a job and commit their resources to a particular course of action. During the time between (1) and (2), when they consider that they have "successful careers," they employ different mental perspectives toward realizing their objectives. Some will view the initial job choice as the first step toward the goal, while others will take the new job as the first piece of evidence that their careers are already successful. In one sense the difference is simply in the perspective you take; in another sense the difference in perspective makes all the difference.

Like leadership, structure is another element of organization needing to treat time as a resource rather than as a constraint. Why are structures always lagging behind? Why can't they exist in real-time? And what would real-time structures look like? We have to be theoretical here, because the applications don't exist yet; we're only at knowing that we don't know.

Industrial models of organization largely exist in three spatial dimensions only. The *first* spatial dimension of organization structure is the span of control. How many people can reasonably report to one boss? Despite any number of variations, the industrial community seems to have reached a consensus around the number seven. When the number of people gets beyond this point, then the *second* dimension enters into the structure. An additional layer is placed between the boss and the seven-plus employees, thereby lengthening the chain of command, but maintaining the idealized span of control. This is the beginning of hierarchy, the pivotal dimension of industrial organization structure. The *third* dimension is introduced by extension in geographic space, where the simpler structure basically repeats itself in different locations. The first three dimensions are spatial, like width, length, and breadth; the fourth dimension is a nonspatial continuum.

How does the *fourth* dimension, time, get factored into the structural design of a corporation? Texas Instruments (TI) is the only company I know of that attempted to include time as a true dimension in its organization structure. But it did this only in the planning function, imaging its business at varying points in future time, generally measured yearly. Different product/market mixes could be identified as they were intended to evolve, and appropriate structures could be designed for each time period. The notion here is that management can more readily ease the organization into real-time structures when it envisions the appropriate forms as part of its planning process and makes changes in modest, but almost continual, increments. TI had the conceptual understanding. Their performance, and the lack of imitators of their fourth-dimension structuring, however, shows that we are still a long way from real-time structures.

Time as an intrinsic dimension of organization means creating real-time structures; structures that change continually in tiny increments, not in large static quantum jumps. Each change is so minute that the overall effect is one of a structure in constant, seamless motion. By contrast, industrial structures are like still photographs. Today, we are learning to assemble a series of these still photographs, and put them on a penny arcade pinwheel, making them begin to move through time. We are not yet able to extend the metaphor, however, to view structures as "movies." This would require a contextual shift.

Any Place

Producer: Your place or mine?

Consumer: If you're going to argue, forget it.

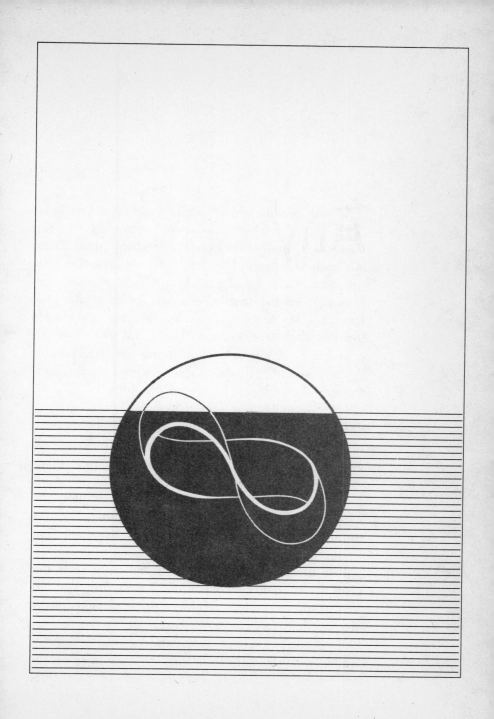

Early generations of products are often big, bulky things. Early clocks did not fit on wrists. The first radios and televisions were ungainly sized boxes, and early computers filled entire rooms. The first electronic calculator was also too large to be convenient. Dr. Wang placed the small display on top of a table, the large core in unused space under the table, and created the first desktop calculator. Today, we carry business card-size electronic calculators in our wallets. The only thing preventing calculators from getting even smaller is the size of our fingers pressing the buttons.

In the new economy, miniaturization of products often involves microspace. The new technologies, built around lasers, fiber optics, genetic engineering, silicon, and artificial intelligence all pack more micromatter into less microspace than did the industrial technologies. Much of this micromatter is really subatomic, electrical impulses that represent information which is transmitted through space as data, words, images, and sound. It is the combination of speed and complexity that gives the computer its phenomenal capability, and neither would have been possible if the constraints of microspace had not been first transcended. Today, you can line up one to two thousand transistors on the head of a pin. Since the early 1970s new generations of denser, more powerful chips have appeared about every three years. Engineers are now shrinking chip circuits to the one-micron level, and below. One micron is one-millionth of a meter, one-hundredth the diameter of a human hair, and one-tenth the diameter of a human blood cell; it is about the length of fingernail you can grow in 45 minutes. In one-micron chips, each layer is about 50 atoms thick. Intel spent $100 million in development costs so that its powerful 80386 microprocessor chip can cram 275,000 transistors and a million electronic components onto a quarter-inch square of smelted sand.

Miniaturizing a circuit on a chip means that the signal with this information can cross the circuit from side to side in much less time. The circuit can therefore run at a much higher speed and conduct many more calculations within a given unit of time. Miniaturization also permits placement of a larger number of logic elements on the chip, thus permitting the machine to handle more complex operations in the same space.

Two additional benefits of miniaturization are reliability and cost reduction. Old vacuum tubes burned out. With solid-state technology used in semiconductor chips, if they survive the heating and cooling processes during the first month of use, they will essentially outlast the life cycle of the product they occupy. Miniaturization also decreases costs tremendously. What used to be many components needing further assembly before appearing as a finished product are now a single component made in a single process. The microprocessor in a personal computer is an example, replacing many different components in older computers. The technological development of fractional horsepower in electrical motors, for example, made possible the development of a slew of consumer products, such as home dishwashers, clothes washers and dryers, vacuum cleaners, and electric can openers.

Electromagnetic motors, however, are bulky and do

not run smoothly at slow speeds. To address this limitation, Japan's Matsushita Electric Industrial Company is developing an ultrasonic motor that weighs less than 2 ounces and fits in the palm of your hand. They intend to use the motor in factory robots, auto parts, and video equipment.

Simple consumer applications follow. Sony Corporation has encoded the four-volume Tokyo yellow pages on about a quarter of a 3.5-inch compact disk. In most cities, a list of sushi restaurants within a five-block radius of wherever you are can be produced in about one second (in Tokyo most of the little streets have no names, and you would be restricted to the name of the ward, section, and neighborhood).

The technological ability to transform micromatter, by compacting it in space, is very much at the heart of the new economy: transforming time, space, and mass to be more useful to people. The cornerstone of the transformation lies in seeing them as resources rather than roadblocks. Thus, *when a spatial limitation is reached, rather than viewing it as a constraint, space needs to be redefined so that it can accommodate the new need.*

Redefining space need not be limited to width, length, or height. Codex, a Motorola subsidiary, for example, juggles 8-dimensional space in a high-priced modem that is sixteen times faster than modems used with common personal computers. As I noted in the beginning of the book, mathematicians at AT&T Bell Labs are working on modems conceived in 72-dimensional space. On an abstract plane, space can be n-dimensional. Think of the esthetic, economic, and social dimensions of the space in your home.

The video revolution is producing dream screens for TV and computers that are flat, superclear, and wide as a wall. The direct addition of another spatial dimension to such products is an even more radical development. Most readers have seen computer graphic drawings of the automobiles and other products in magazines. They are line drawings, like architectural sketches, that simulate three dimensions. These simulations are even more realistic on television, as the viewer's perspective is rotated by the computer's eye view. In all cases, however, these are never more than 3-dimensional simulations captured on 2-dimensional screens. In fact, most of the products we have created to capture information utilize essentially 2-dimensional space.

Records, film, tapes, screens, and disks are all 2-dimensional formats for capturing and displaying information. The spaces they occupy, for all practical purposes, are flat planes. Technologists I have spoken with say there is little in the way to prevent manufacturers from "folding" these 2-dimensional products up into 3-dimensional ones, creating records, film, tapes, screens and disks that are shaped like cubes, with the encoded information read a layer at a time. Ampower Technologies Inc. of Fairfield, New Jersey, is developing a rotating lens that permits the image on a cathode-ray tube to actually be projected about 18 inches into space in front of the CRT. This is the precursor of true 3-D television, and a whole lot more. In terms of the number of spatial dimensions, this is the difference between photography and holography – the technological precursor for holistic modeling in other fields.

In physics, density means the ratio of an object's mass to its volume. Applied to products, density should be used to mean the ratio of product size to its value. Shrinking the size of a product without otherwise changing it, in and of itself, creates greater value. Enhancing the value at the same time is even better.

Reinforced concrete is an enhancement of ordinary concrete because it packs more value into the same space. Infection-resistant strains of crops are similarly enhanced. A dedicated word processor is an enhancement over a typewriter of the same size, and, when that word processor can also handle data, its value increases even further. These are all examples of product enhancement. Enhancement of services is even more straightforward, because additional services often can be added to the same space quite easily. When an accountant adds tax planning to her services, or a restaurant adds atmosphere and a maitre d', they are increasing their value, not their space.

Another example of enhancement is in a communication system first developed by AT&T as TASI (Time Assignment Speed Interpolation), and more recently as Burst Switching, developed by GTE. The marriage of computers and telecommunications is placing a strain on telephone networks, switching, and transmission facilities. Without technological enhancements, current networks will be unable to cope with the anticipated traffic of the 1990s. The enhancements come from parallel developments in microelectronics, silence detection, speech, voice compression, and signal processing. Burst Switching triples present communication capacity, and processes calls twenty times faster, by packing more value into the same space.

It can do this because speech is two-thirds silence, interspersed with bursts of sound from 0.1 to 1.5 seconds long. Instead of locking up an entire line during a telephone conversation, the lines on the new system need be dedicated only during the very brief duration of voice bursts. Data and other "bursty" conversations can be shoehorned in during the other two-thirds of the time, effectively tripling transmission capacity by packing more into the same space.

Miniaturize something enough and space becomes

even more of a resource and less of a restraint: the product becomes portable, easily carried through space. The importance of miniaturization is that later generations of a product are so compact they can go with you wherever you are. Imagine two products that are identical in every way but one: the consumer has to go to product A in order to use it, product B moves around with the consumer. Which would you choose? Voilà, the wristwatch.

How many products can you think of that started out stationary and were then made portable? Any product that is information based, whether in the form of sound, image, words, or data, can probably be adapted to any time, any place delivery. Not many people carry a dictionary, encyclopedia, or other reference texts around with them, but if collectively they were no larger than a 3.5-inch wafer (CD-ROM), the market would certainly be there. I now sit next to executives on airplanes who, having done their spreadsheets on computer screens that lie on their lap, send them to their office via the phone that hangs in the aisle. Their work is accomplished before they arrive!

Even if the product is too large to lug around, it can still deliver the service to you. You don't have to go to it to get the service. Pitney Bowes' Postage By Phone, for example, allows mailroom managers to call them, punch their code number in, and get their postage meter refilled by phone. No more getting to the Post Office before it closes, going out in the rain, or waiting in line.

Home health care is a $14 billion-a-year service that is based on shifting the place of delivery from the producers' to the consumers' domain. In our grandparents' time, most health care took place at home. Then, with technological advances, the locus of health care shifted to the hospital. Mainly for reasons of cost, and secondarily for customer comfort, medical care is coming home again. More than 7,000 commercial agencies offer home care services. About half the hospitals in the U.S. are in the business, together with companies as diverse as Avon and H&R Block. A 1985 survey showed that nine out of ten Blue Cross Blue Shield plans offer some home care coverage, up from 5 percent a decade ago. Moving delivery systems out of the producers' space to where the client is can help build more viable institutions.

The new AT&T, in its approach to teleconferencing, missed the importance of carrying out service in the customers' space. AT&T lacked a marketing culture, and when it built lavish teleconferencing centers in the business heart of a dozen U.S. cities, it expected travel-weary executives to prefer a taxi ride downtown to a plane ride across country. However, the centers languished and closed. Executives did prefer teleconferencing to travel, but only when it was done in their own offices.

When televisions first came out, the motion picture industry first ignored them and then were tremendously threatened by them. It took the movie industry decades to realize that television provided them with an entire generation of additional growth. Now, as VCRs have entered the market, movies are reacting in similar ways. Worried once again about the threat, they are positioning themselves defensively through law suits to capture what they see as lost revenues. It will probably take them another decade to see the VCR as a boon to their industry. In both the TV and VCR businesses, the consumers opted for products that they could have in their home over the movie house.

Citicorp, in conjunction with the Panasonic unit of Matsushita Electric, has developed a palm-sized personal computer terminal that lets people anywhere do business with them without setting foot in a bank. The portable unit is about the size of a large hand-held calculator, and plugs easily into a telephone. Anyone using the terminal, after automatically dialing a Citibank computer, can access their savings, checking, credit card, credit line, and other lines; they can check their balances, transfer money between accounts, look to see if checks have cleared, and eventually pay bills. And, they can do all this without the personal computer currently required in electronic home banking systems.

Nonbank banks, such as Ford and Sears, are considering franchising stand-sized bank service huts in shopping centers. Citicorp's consumer strategy goes even further: to deliver full banking services inexpensively to large numbers of people without regard to geography or physical facilities. Its success depends heavily on technological enablements, such as the hand-held terminal. These developments also bring people closer to nationwide banking, by skirting such federal laws as the one that prohibits deposit-taking outside the consumer's home state.

The ultimate in portability is more than a product easily carried around because it is small; it is also a product that can move around with the users wherever they are because it doesn't have to plug into a power source.

In 1876, Bell invented an instrument for conveying speech over distances by converting sound into electrical impulses sent through a wire. For over a century large numbers of people could talk with each other – as long as they were connected by wire. Now, a physically connecting thread through space is no longer needed. Wireless telephones are popular, although people are still tied to within a short radius of the wiring system. With the coming of cellular radio telephones, however, even this constraint is eased. Wireless cellular telephones can be moved around in a car, but they are bulky things. Now Technophone Ltd., in Woking, England, is making a 15-ounce, battery-powered phone that can be carried in any briefcase or purse.

Speaker phones further ease the space constraint, and we no longer have to stay so close to the transmitter and the receiver in order to communicate. So far, the appeal of speaker phones has been to the business market, but they could be repositioned and made cheaper for kitchen and den. Increasingly, we are free to move about as we communicate with each other wherever we are. Large numbers of people around the world will be talking with each other from and to any place they happen to be.

The portable radio was another technological advance that eliminated physical connectedness. The Sony Walkman enhanced the product further by combining quality sound with miniaturization. Panasonic patented TriTex circuitry, which reduces the stereo radio to the size of a postage stamp which is attached to the ear clip of the head set. Perhaps the next development will be a microradio lodged in the back of an eyeglass frame.

Even the most mundane services are learning to transform time and space, and use them for competitive advantage. Domino's, for example, sends over $1 billion worth of pizzas from 3,400 stores to home deliveries within thirty minutes of receipt of the order. The customer dials a local home delivery service number. The order is taken at a computer terminal in less than thirty seconds and is sent electronically to the closest satellite bake shop for preparation, then into insulated plastic bags, and onto radio-dispatched delivery vans. Pizza Hut promises their pizzas free in half their markets if they fail to get it there in less than thirty minutes. In Tokyo, eight pizza chains opened home delivery services in

1986, and local newspapers report on the number of minutes by which rivals shorten their promised delivery.

The practical message here is that consumers are more likely to use the products they can have with them wherever they are than competing products they have to go to in order to use. Making products and services less fixed in space increases their competitive edge. If customers have to come to you in order to use your products, or avail themselves of your services, figure out how to make your deliverables available to them in their place, rather than yours. Hoover and Avon did this with door-to-door sales many years ago, technological advances have done this by molding the deliverable itself, and yet another shift is making this happen in manufacturing.

In an industrial economy, goods are produced in the physical space of the manufacturer. From there they are distributed, often through intermediaries, into the hands of the consumers. In the new economy, the end of the manufacturing chain of goods and services increasingly will be produced by consumers, in their own physical space. Probably the earliest example of this in the new economy was do-it-yourself, and the basic motivation was cost-saving. Since then, other instances show different motivations.

Polaroid instant photography is another example. Processing, the last step in the manufacturing chain, takes place in the hands of the consumer. The basic motivation here is time-saving, instant gratification. Modularization is another manifestation of the same principle. Instead of buying a self-contained hi-fi system, for example, the consumer buys one manufacturer's speakers, another's turntable, and a third's amplifier. The consumer puts together the final customized system. In banking, to give a fourth example, the physical structure moved over a period of time from the main office into branches, then got physically closer with drive-up tellers, got closer still through ATMs, and closer still by eliminating both time and place in point-of-sale debiting. Each of these spatial shifts gets closer and closer to the physical space of the consumer.

The ultimate shift occurs when the producing organization moves *into* the physical space of the consuming unit, and the consumers take over the producer's actions in their own space. This is what occurs in electronic home banking. The consumers sit in their own space (home) and electronically create their own bank transactions. When a bank was a building, we knew where it was. When the ATM became the bank (it accepted deposits, made withdrawals, and handled other transactions as well), the space of the banking organization became harder to identify. When the bank exists in the software of the customer's computer, where is the physical bank?

Another example, and one that has been around for many years, is adding water to dehydrated or condensed products. For years, Coca Cola was sold at soda fountains, where the syrup was mixed with water in the retailer's space. Recently, Coca Cola executives debated whether to build home dispensers into refrigerator doors, next to the ice-cube and ice-water dispensers. Since water takes up the largest volume of the product, they would be reducing the product space tremendously, cutting costs in bottling and distributing, enhancing the value of the refrigerator, involving the consumer in the end stages of manufacture, and probably increasing consumption. Sodastream, a U.K. company, had a variation on the idea; they sold a cartridge that would add carbonation to any liquid, a very cheap way to make your own sodas.

Computer software is a major extension of this logic. The consumers use the software to create the final applications, and hence products, themselves. Similarly, computer-aided design lets the customers design their own specifications. The same lesson that is true for products exists for services. Working on their home computers, customers can plan their vacations, make their bookings, purchase airline tickets, and negotiate the particulars of the vacation loan – all steps in the final creation of a travel service package, conducted in the buyer's space. In the future, you need not go to a store to buy a software package, but can order it up by phone, and have it sent electronically to your computer, where you will manufacture it on a disk. This can have an anti-copy command built in, as well as the number of days or times it is usable.

Shifting the determination of a product's final configuration downstream, into the space of the consumer, has very practical consequences. Consumers who create and control the manufacture of their goods and services are likely to consume more than those who do not. The lesson is analogous to people's behavior at a buffet table. They always fill their plate more when they do it themselves than when a waiter serves them. The practical message: finding ways to move the end of production downstream, into the space of the consumer, will result in renewed growth for the product.

In sum, dehydrated food, the drive-up window, shopping at home, "smart" buildings, manufactured housing, and factories in outer space would not have been possible without the producers first shifting their sense of the space their products need to occupy. Practitioners should create products and services that are not bound by traditional conceptions of space and its constraints. Product development people, particularly those working with mature products, would do well to rethink all aspects of their products' space – size itself, size relative to value, where it's produced, and where it's consumed.

Two products can't occupy the same space at the same time, but perhaps two intangible needs can. The new economy has defined markets in this intangible way, resulting in markets that don't have spatial determinants.

In the postindustrial economy, a market is no longer a physical place for buying and selling but an identified *need*, a rather intangible criterion. The people who have the need are the actual or potential customers, and those who fulfill the need are the producers and distributors. The marketplace is the space where the function is performed. Part of the concreteness of markets, thought of in the industrial sense, is producers and consumers located in specific physical places, which is no longer always true. The meaning of market "place" is being fundamentally transformed for both the seller and the buyer.

Where, for example, is the stock market? Is it 11 Wall Street, the New York Stock Exchange, with its marble columns and traders running around the floor exchanging slips of paper? No. The competition for exchanges in the new economy does not occur in a "place." The National Association of Securities Dealers, founded in 1971, is providing some of the strongest competition with their automatic quotations system (NASDAQ). In 1987 they trade an average of 155 million shares a day of over-the-counter stock for smaller companies. Although the size of companies whose stock they trade is smaller, they list more than three times as many companies as NYSE, and their list is growing whereas NYSE listings have declined. Unlike NYSE, NASDAQ does not have a specific market "place." Instead, it has a network of dealers who flash each other quotes on 3,000 computer screens over a six-million-square-mile trading floor. A dealer who has an order to sell 1,000 shares of Convergent Technologies, for example, scans the NASDAQ computer for the highest price and, together with a phone and modem, completes the transaction.

Stock market trading is moving into an any time, any place world. The market exists not in one place but rather in every place and, because of the London/New York/Tokyo axis, it also is moving to 24-hour-a-day trading. The Chicago Board of Trade, for example, opened evening trading sessions in May 1987. This can only be done with totally computerized systems. The meaning of the stockmarketplace is completing a fundamental transformation begun decades ago. Merrill Lynch began the trend of bringing securities trading action physically closer to the customer with the phrase, "We bring Wall Street to Main Street." The process is just moving toward its natural conclusion: conducting these activities whenever and wherever they are needed, without limitations of time and place.

Marketers are also redefining what they consider to be the relevant space of their customers. Some of these include a market share of stomachs, voice, and mind.

I was talking about this in a management session with Coca-Cola bottlers when they told me that they work seriously in terms of "share of stomach." Stomachs have only so much room, a portion of which gets taken up with liquids. America's drinking tastes are changing: moving away from hard liquors and even wines to non-alcoholic beverages; away from caffeine; and toward sweeteners other than sugar. If you don't want to lose your share of America's stomachs, you have to change with America's changing tastes. Before Coca-Cola's major marketing split between the old and new formulas, a 1-percent share of brand Coca-Cola represented around 77 million cases. Even the slightest shift in available stomach space makes a big difference to them.

In heavily advertised businesses, some executives correlate their products' success with what they call "share of voice." Share of voice, to them, is the relationship between the percent of total expenditures for communications about the product and the public as a percent of that product's market share. If a product has a 10-percent market share, and only 7-percent communications share, then its share of voice is considered less than it should be. The communication is also taking place in previously unthought-of or unused spaces. The cost of prime-time TV ads skyrockets to about $100,000 ($600,000 during the Super Bowl) for thirty seconds, even though networks' share of total advertising has dropped 15 percent in a decade. This is causing ads to pop up on parking meters, on shopping carts, on huge screens at sports stadiums, on products slipped into movies, and even on clothing, as Coca-Cola has done.

Despite its unusual location, stomachs are in a pretty definite place. Less definite, though just as serious, however, is the marketing concept "share of mind." Retailers carve up their markets along many dimensions, including income, life style, and location. Even within a defined segment, however, there is still a competitive battle. When similar products can be purchased through many end points, it is difficult to distinguish one product from its competition. Merchandisers resolve this by emphasizing the institutional image rather than the product image.

Value, then, is not in the space of the product so much as in the space the product is in. Retailers understand that their customers often come to their store because of the store's image as much as because they carry the desired article. If you need an ashtray or a pair of underpants, you can get them pretty much anywhere.

Because these are near-commodities, the store they are bought at creates much of the perceived value. Those items bought at Neiman-Marcus or Bloomingdale's are not the same as the ones at the corner store. Upscale stores like these want to capture a large share of your imagination, self-image, and other aspects of your mind. They want their store to be very present in your mind, and probably in the brain's right hemisphere.

What these examples show is that the meaning of market "place" is shifting away from the physical space of industrial economies toward less-fixed locations in the space of the new economy. Managers should take another look at their markets and also define them in non-spatial, non-locational terms. The traditional market definitions will not disappear, they will be complemented by new ones. Product managers will fight not only for shelf space but also for mind space, having customers thinking of their products when they go shopping.

Transformations of time and place will affect both market theory and practice. When markets expand from nine-to-five to around-the-clock, has the market grown? And how has the market opportunity developed? Is it the same market, now served differently, or does it tap into a new and different market? Which markets are the most, and least, likely to move in the "any time, any place" direction? They will not move there in a random way. Young professionals probably need their market developed in these directions more than blue collar workers or retired people. What kinds of costs are involved in servicing any time, any place markets? Are the costs based on hours and locations, or on technology and organization complexity?

These are questions about market theory, yet they also raise questions of a very practical nature. For example, how does a company get its marketing people always to be thinking about fulfilling market needs in whenever and wherever terms? When a need in the market is identified correctly, as much creativity should go into – and as much value-added should come out of – determining *when* and *where* to meet the need as goes into *how* to meet it.

In the same way that the meanings of markets are changing, so too are the functioning of markets, and some marketing functions. In a complex industrial economy, for example, a tremendous amount of space in the marketplace is taken up by intermediation. Finding ways to disintermediate is a way to create space for new opportunities.

A marriage broker is an intermediary. When young people find their own partners, however, they have disintermediated the marriage broker. Similarly, a stock broker is an intermediary, acting as buying and selling agent between the customer and the company issuing the stock. The broker further uses the intermediation of a stock exchange. When people bypass the broker, they are disintermediating a go-between.

In simple economies people carry out all steps in the chain that creates economic value. They secure their own resources, use them to produce their own goods, and consume those goods themselves. This is the subsistence farmer. When the farmer supplies food to the city dweller, he is an intermediary between the natural resources and the urbanite. The more specialized the jobs become, the more they are intermediary between two other economic activities, one further upstream nearer the raw materials, and the other further downstream closer to consuming the deliverable. The more complex the economy, the more intermediaries there are in the chain, each supposedly adding value by their contribution along the way.

Banks intermediate our money; retail stores intermediate our furniture, clothing, and endless number of goods; television and movies intermediate our entertainment; utilities intermediate our forms of power and energies used; trains, planes, and buses intermediate our movement; schools and hospitals intermediate our learning and healing; and government intermediates a whole range of services, including our protection.

Service economies are reacting and adjusting to having too many intermediaries. Every intermediation adds costs that are passed on to the consumer. In some industries, such as food, clothing, and cosmetics, intermediary activities are the largest component in the final price of the product. Eliminating intermediation is one of the key ways to being the least-cost producer. The more intermediary a job or business is, the more vulnerable it is to disintermediation. Viewed from the industrial context, disintermediation creates negative space in the form of lost jobs. From the new economic context, it creates positive space, by reducing the value-added chain to its most efficient number.

In the cut-flower business, this has meant great savings for the consumer, greater competition for the independent florist, and a bonanza for flagging supermarket chains. Eighty percent of the nation's major supermarkets now stock flowers, more than twice as many as in 1974. Together with discount stores and department stores, they now represent an almost 20-percent share of the retail cut-flower market. Big chains bypass flower wholesalers on whom small, independent florists rely by buying in bulk directly from flower producers and storing them temporarily in their own warehouses. Kroger, for example, the Cincinnati-based supermarket chain, buys 80 percent of its flowers this way.

Disintermediation can, itself, become a business. Mail-order is the disintermediating business, and one of the fastest growing sectors, in retailing. Videocasette recorders are disintermediating television and movie theaters. Health maintenance organizations (HMOs) are a disintermediating agent in health care. Generic goods are a disintermediating agent in advertising. Do-it-yourself has disintermediated a host of fields. General Motors Acceptance Corporation (GMAC) is a thriving business that disintermediates banks by financing the purchase of its automobiles directly with the consumer. Corporations further disintermediate banks by issuing their own commercial paper, another new business.

Disintermediation not only occurs macro-economically, between organizations, it also goes on within companies. Whenever a report is scrubbed, some intermediary activities are cancelled along with it. Staff are the ultimate internal intermediaries, and their reduction is synonymous with disintermediation. Even without reductions, however, staff are disintermediating themselves and enhancing the companies' overall economic value at the same time. When staffs charge the line for services that add value, they are, in effect, forcing the line to reduce uneconomic overhead that often comes from intermediation.

What kind of agent is most likely to create disintermediation? The answer is small business, where most people do for themselves what in large businesses they would have specialists do for them. Two out of every three new jobs created in the past ten years have been in businesses with less than twenty employees, whereas the Fortune 500 created no additional jobs during that same decade. While this is testament to the shrinking industrial economy and the burgeoning entrepreneurial economy, it is an oversimplification to say that the former is sick and the latter is healthy. More accurately, the new economy is transforming how economic activities are arranged in the marketplace, in both small and large organizations. The disappearance of older arrangements is making space for the emergence of new market configurations.

The need to rethink relevant market space is nowhere more important than in terms of distribution. Distribution, one of the major activities in marketing, is the movement of goods and services from producer to consumer. Traditionally, this includes handling, storage, and transport, and excludes sales. Defined more broadly, it is the application of people, technology, and outlets to acquire customers through advertising, promotion, and sales, and to maintain them through service, revenue collection, and delivery on their transactions.

Distributors are often candidates for disintermediation, but technological developments have created great untapped leverage in the distribution part of the value-added chain. This becomes particularly important in service businesses where distribution represents between 45 and 80 percent of operating costs.[1]

By 2001, microcomputers will be as common as pencils. They will be commodities long before then, and, unlike pencils, they will be connected with one another in networks. To the hardware vendors this means that profits will come from further down the value-added chain, particularly from distribution, rather than from the manufacturing side of their businesses. Deregulation and changing consumer preferences have also had a marked impact on distribution. These shifts are already significant in distribution activities in retail gasoline, airlines, motion pictures, photoprocessing, trucking, and financial services.

In retail gasoline, with margins guaranteed under government regulation, the singular goal was to increase volume. The result was dense distribution networks – "a station on every corner." Revocation of the oil depletion allowance, the OPEC embargo, and resulting price increases had dramatic effect: demand dropped, margins declined, stations closed, point-of-sale systems emerged, and self-service and convenience stores at gas stations grew rapidly. ARCO and Shell capitalized on the new conditions by reconfiguring their distribution networks.

ARCO targeted customers wanting low-cost, branded gasoline. It then created distribution efficiencies by

- Reducing activity from forty-eight to seventeen states and eliminating 60 percent of its outlets.
- Converting more stations to self-service than any other major brand; thus lowering costs even further by shifting part of the distribution activities to the customer. (Note the parallel to the shift in manufacturing location discussed on page 53).
- Eliminating its credit card and passing on the 3.5 cents-per-gallon savings to customers.
- Shifting incremental revenues from repairs and parts to convenience store items.

The average volume per station quintupled from 1973 to 1983. ARCO radically transformed its sense of relevant market space and the strategy worked.

Shell responded to the changing environment with a different strategy, aimed at customers who wanted competitively priced gasoline *and* credit. They began accepting MasterCard and Visa, as well as any other major oil company credit card at no fee – when most companies were charging 3 percent. They are leading the move toward low-cost, unmanned gasoline stations, and are in the forefront of point-of-sale technology applications. As an example of the latter, Shell adapted J.C. Penney's nationwide system for credit authorization to reduce credit fraud and float costs. Their successful strategy is gaining leverage by reconfiguring the place of distribution in their retail business.

Texaco, by contrast, held on too long to their strategy of serving all market segments in all fifty states. Under the new environment, they were trying to serve too much market space and failed to recognize that higher costs meant local stations had to increase volumes to break even. With an inappropriate distribution system, their market share dropped nearly a third between 1973 and 1983, from 8 to 5.8 percent.

The airline industry is another example of the changing place and shape of distribution. Also hit with deregulation, the number of carriers increased, but recession kept demand flat. The result was fare wars, and costs exceeded revenues in 1980–1981. Major airlines shifted to large cities, making room for commuter lines, and also expanded the number of their fare offerings. People Express, a new entry at the time, shaped its distribution network around a low-cost strategy: extra fees for services such as baggage checking, advance ticketing, or on-board meals. They quickly won a lot of market space with the no-frills approach, although they later lost it through financial mismanagement. American Airlines won as an incumbent by combining quality and full service at a moderate price. Their distribution systems were changed to reinforce their strategy: using three hubs (Dallas, Chicago, and Denver) instead of one allowed more flights with the same number of airplanes; their SABRE reservations system locks in 7,000 travel agents; and they were the first to offer a frequent flier program.

These winners in the above examples all chose to serve selected markets, to differentiate themselves from their competitors through their distribution systems, and to manage the cost/revenue relationship. The lesson is not restricted to their industries.

From a market perspective, customers get value from

the distribution system as well as from the product or service itself. The two together determine the value of the offering, and different customers need different offerings. The distribution system can add a great deal or only a little value to a deliverable that may be either customized or a commodity. Generally, the distribution systems for commodities add little value to the deliverable, whereas those for customized deliverables add quite a bit.

In the restaurant business, for example, a fast-food outlet simplifies and automates the preparation and service, where the product is understood, and little information or advice is needed. An elegant gourmet restaurant, on the other hand, will lavish advice and attention on the customer, thus combining a high value-added distribution system with a customized product. Gourmet and fast-food customers are in different market segments, yet both are selective shoppers whose distribution expectations are in line with the complexity of the product they buy. Volume is greatest, and costs most aligned, along this diagonal of selective shoppers.

Another market segment are the service sensitive customers who demand special attention even for commodity transactions. These are the few who will pay for the privilege of going to a human teller instead of to a teller machine. The opposite segment are self-service oriented do-it-yourselfers. These are the types that use discount brokers and order IRAs by mail; they are yuppies who will buy gourmet fast food. As their experience and knowledge grow, they move toward saving by expecting less from the distribution system though not from the product.

Bantam Books' "Instant Book" combines real-time production with innovations in the distribution system. By tying together production functions such as typesetting, and on-line order processing, telemarketing, inventory, and financial services, Bantam can compress a one-year publishing cycle into three days. When the U.S. Olympic hockey team beat the Russians on a Sunday night, Bantam had *Miracle on Ice* in bookstores that Thursday morning. Not all books need be instant ones, but most authors, publishers, booksellers, and readers would appreciate major compressions of the time lag, and a large part of this lag is in the distribution end of the chain. Four years ago it took four to six weeks for a bookstore to get a special-order book for a customer. Today it takes an average of four days, and in a few years we can expect this wait to decrease to overnight.

Most providers of products and services still rely heavily on a single dominant distribution channel, even though their offerings vary and their clients respond to multiple channels. New competition arrives in the spaces that are not covered, serving the same niche by improving on the distribution mechanisms. The relevant "space" here is the selected product-market niche matched with the appropriate distribution activities. Thought of in purely physical terms, distribution was more a constraint than a resource; redefining this space, along non-physical dimensions, distribution becomes a potential competitive advantage.

Transformations of space began with new scientific understanding, then technology transformed our sense of space in products, services, and markets. In an "any time, any place" economy, spatial transformations will also occur in our organizations. This is becoming obvious in the varied and important new meanings of size.

In the industrial pantheon, growth has always meant large size. As long as return on investment was above minimal respectability, then the true measure of corporate success was revenues, not profits. That most-quoted yardstick of importance, the Fortune 500 list, is an industrial ranking of manufacturing companies based on size, not performance. America produced and blessed giant corporations. In recent years, however, the value of these giants has come into question, and other yardsticks – based around value and quality rather than size – have been introduced.

Corporate raiders have argued that the whole is worth less than the sum of the parts of many large corporations.

The raiders argue that their target companies take up more space than their contribution warrants, dissipating the nation's economic energies.

Businesses inevitably mature, and as they do they may throw off more cash than they can profitably re-absorb. Management can react to the absence of opportunities to reinvest internally, either by returning more of their cash flow to shareholders in the form of larger dividends, or by investing externally. The majority chose the latter route, diversifying into businesses that they may or may not be familiar with. Because it is very difficult to gain a larger share of a mature market, managers do not often use the lionized cash cows to buy more grazing space. Instead, they use the money to finance new and often unrelated ventures. Their track record is not very good. Their companies take up more and more space, less and less efficiently. When this happens, it deteriorates the public welfare as well as the business climate.

The oil industry is a perfect example. Many oil-producing nations are building their own refineries in a world market that already has more capacity than it uses. The industry is very capital intensive and has an enormous cash flow, which it cannot profitably reinvest internally. As in most American business, the oil giants solved their dilemma about what to do with their excess funds by using them to occupy more space, not less. In other words, they diversified. Some of them used the logic that they should go into businesses with similar capital structures, like minerals and metals. Others used the logic that they should go into businesses with countervailing capital structures, high-growth fields that gobbled up cash, like office equipment. None did a very good job, and, for a long time, none chose to return more money to their shareholders and shrink back – to take up less, though more economically efficient, space.

Enter the raider. And enter the debate as to whether he is good for America or only out for himself. When an oil company's proven reserves in the ground are worth more than the entire company's value in the stock market, the whole is worth less than the sum of the parts. Buying the company and selling off the parts is probably a good thing to do if the stock is undervalued, because managers are not maximizing the most efficient use of their resources. It may not be a good thing to do if the stock market is undervalued, because the financial markets are not pricing existing assets correctly. Whichever the case, however, and whether the raiders are ruinous spoilers or shrewd investors, they are clearly causing corporations to scale down. Oil companies, and many others, are buying back their stock in tens of billions of dollars. This provides for larger dividends and price of the remaining shares, and causes the funds to flow from the oil industry into more profitable places. Not all of this is good, because some companies incur huge debt in stock buybacks, whose only purpose is to protect the existing, underperforming management.

In either case, size is no longer a guarantee of security. Mobil Corporation produced $56 billion dollars of revenue in 1984. Only 18 percent of those revenues came from petroleum exploration and production, whereas about 60 percent came from refining and marketing; the rest came from retailing, and a little from chemicals and packaging. Mobil's profit picture, however, is quite different: 91 percent of its after-tax earnings came from exploration and production, and only 9 percent came from everything else. These figures were even more skewed in 1985 due to Mobil's restructuring of Montgomery Ward.

Like the dinosaur who used most of its brain capacity simply for internal maintenance, Mobil spends most of its resources running parts that do not contribute to its total health. Integration had been the common wisdom in the oil patch, but disintegration is the more likely future. If Mobil scaled back to what it does best, and to what the stock market appreciates most — finding oil and gas — then a $10 billion mini-Mobil would produce 90 percent of the same profits, with a small fraction of the resources it currently uses.

Many businesses operate with what is known as the 80–20 rule: 80 percent of their business comes from 20 percent of their customers. Serving the additional bulk of customers is not terribly rewarding. To do so, however, means growing a large organization that pushes up sales far more than profits. Its contribution to the economy, relative to the space it occupies, is not very productive. This does not mean that such a company can produce 80 percent of its profits by using only 20 percent of its resources, while redeploying the remaining majority elsewhere. Nevertheless, it does mean that a very large portion of a company's resources are not being used productively, and that many firms can scale back considerably with only a minor effect on the bottom line, at worst, and probably a more salutary effect in the larger number of cases.

Few companies can survive for very long when their return on capital is less than their cost of capital, and most companies cannot provide a return on equity that is greater than their cost of capital. This caused companies to turn to off-balance sheet financing, as in fee-based income, to generate income without acquiring assets. The good news in all of this is that you can get your company to grow 100 percent. The bad news, however, is that you can never get your investment back. What matters more than size, therefore, is the quality of the assets in relation to their earnings.

These financial and economic changes have important implications for organizational change. New models will manage *space* as a key resource in transforming their products and services, markets, and organizations. Downsizing, for example, is a redefinition of how large a space is optimum for a company to occupy. In an economy measured by growth and a competitive arena rewarded by market share, revenues – and hence large-size organizations – are an appropriate yardstick for determining optimum space. In an environment judged by the rate of return on assets at risk, however, organizations that have fewer and smarter people creating greater capital returns are more highly valued. From this newer perspective, proper organization space is not measured by size, but by size relative to return. The economies of scale from a production point of view are balanced or replaced by a concern for the diseconomies of scale and coordination from the point of view of the space of an organization.

Also, while financial resources distinguish between assets, earnings, and sales, methods of human resource accounting are not yet developed to the point where they distinguish between the asset value and earning contribution of people in contrast with the antiquated measure of body count relative to sales. Human resource accounting along these lines had a brief flourish in the 1970s, but then disappeared. It is not clear whether the failure was due to the methods of counting or to the perceived lack of a real need at the time. If companies would develop such accounting devices, then the meaning of employee size would be very different, and many shrink-ages and layoffs might never become necessary in the first place.

To take one type of accounting device as an example, industrial reward systems often judge the importance of a position in terms of the number of people below, in the reporting hierarchy. Reward systems that are more consistent with economic orientations of the new economy, by contrast, might reward managers regularly for the number of people they can reduce from the reporting hierarchy below, and not only at crunch times. This gets close to a value-added approach as the basis for compensation and reward systems. It requires a new understanding of what the manager is managing – people, finances (sales, profits, growth), machinery, or the value-added of these.

When people adjustments are required, they should first be made in the measurement systems that determine the context for the cutbacks and not just in the head count itself. Chopping people, without first transforming the method for determining how many were needed to begin with, is a short-term palliative that is bound to produce the same crisis over again in a short time. It is like a crash diet but without subsequent change in the type and quantity of food eaten regularly, the amount of exercise, or the number of drinks and cigarettes consumed. The context needs to shift, or else the content – whether it be people or pounds – will reinflate regularly.

Nor can all improvements take place by cooking resources down, like a good French sauce, to their essences. Growth in some businesses requires, not downsizing, but quantum leaps in size. The quantum leap phenomenon is often given the sobriquet "strategic alliances," and two questions about them are the most frequently asked: (1) Are they really any different from the joint ventures produced with an industrial mindset?, and (2) Why do the corporate divorce rates rival those in the general society?

Strategic alliances *are* different from joint ventures, and if they are to succeed in greater numbers than their predecessors, they must be viewed holistically. Here are some reasons why:

The joint venture model (based on mechanistics)	The strategic alliance model (based on holistics)
The parent companies are the constituent parts, and the new third company is the resulting whole.	The parents and the third company are only some of the constituent parts of a much larger whole, that is made up of more interrelated parts.
The new whole is smaller than the parts that made it.	The new whole is larger than both corporate parents and their offspring.
The new whole fits into an existing business sector.	The whole is a new business definition, rather than a new entity within an existing business sector.
The new company may be formed for an independent advantage that may be unrelated to the purpose of the parents.	Formed for intrinsically related purposes.
Can be formed by companies of any size.	The whole is so redefined, large, and differentiated that even giant companies are not going it alone.
Stresses the importance of the parts, i.e., the parents and the offspring.	The interrelatedness and interconnectedness of the parents and offspring are the key focus of the alliance.
When the new company succeeds, and especially when it gets to be quite large, it should cease to be a joint venture and should be spun off as a separate whole.	No part, parent or offspring, has viability except in permanent and intrinsic relation to every other part of the new whole.

The merging of computers and communications businesses offers a useful example of the kinds of alliances we are beginning to see. The two major players, IBM and AT&T, for example, are drastically reorganizing their businesses and organizations, and the meaningful space that each of these giants occupies is being transformed as a consequence.

IBM had always been an essentially do-it-yourself kind of company. They avoided reliance on outside low-cost suppliers in favor of their own manufacturing capabilities. They were skeptical of expertise "not-developed-here." They hadn't made a major acquisition in over two decades. Through the 1970s behavior like that was acceptable because of the way computers were used. Computer users sat at terminals connected to giant mainframes; the terminals displayed information and sent instructions to the mainframe, which used its calculating power to implement the commands. The process was centralized, and so was the IBM way of doing things.

Technological developments such as the microprocessor then made it possible to put that calculating power in small computers directly on users' desks. This was cheaper, more flexible and efficient, and less controlling; but the parts of the whole were not linked to one another, and each part had only a limited memory and no quick access to the vast central files in the mainframe. It is this next generation leap, tying together all information-handling parts into a vast global network, that finally caused IBM into alliances which fundamentally altered the type of organizational space it occupied. Growth was no longer going to be a purely internal affair. IBM's organization space was about to expand through a host of new alliances.

Computers and communications are merging into an information handling megabusiness where, as IBM's boss

John Akers says, "everything is connected to everything." The basic pieces of this new whole are *computers*, a *pipeline* to carry the information between the computers, electronic *switchboards* to hook the computers and pipelines together, and the *software* to enable different computers to understand all messages. Putting together the pieces of this new megabusiness has meant putting together a new organization, one that is fundamentally different from the Big Blue monolith and is better characterized as adapting through alliances.

In computers, IBM bought about 20 percent of the semiconductor company Intel, maker of the microprocessor in the PC. Much of the PC is an assembly of parts bought elsewhere. It has even agreed to buy entire computers from Stratus, makers of specialized systems for bankers and brokers, and market them under IBM's label. It began retailing the PC outside – and in competition with – its own sales force, through Sears and Computerland.

For access to the pipeline, it took an initial 18 percent of MCI, AT&T's major competitor in long distance service, the basis for a global voice, data, and words network. For switchboards, it bought Rolm, starting with a 15 percent share and then raising it to full ownership. On the software side we will see a host of alliances, which began with a joint IBM, Sears, and CBS effort to create a nationwide commercial videotext service for home shopping, banking, and information. This was followed immediately by a deal with Merrill Lynch to develop and market an information delivery and office automation system for brokers, banks, businesses, and homes. The alliances will continue to expand.

AT&T is lining up its alliances as well and, because of its pre-breakup domestic nature, it is stressing its need for global space. Whereas IBM had the basic computers and software pieces and lacked the pipeline and switches, the reverse is the case for AT&T. In little over a one-year period, for example, AT&T made arrangements with:

- Six of the top European computer makers to adopt Unix, AT&T's standard for a software operating system.
- Phillips for marketing switching and transmission equipment in Europe.
- Wang to produce compatible computers and document standards.
- Olivetti to buy 40 percent of their company, and hand over their computer business to be managed by them.
- 50 percent interest in a new computer communications company with several prestigious Japanese companies, including Fujitsu, Hitachi, Mitsubishi Electric, Mitsui, Nippon Steel, and Sony. AT&T will build and manage the transoceanic transmission network, and the partners will provide the programs.
- A chip facility and a plant to manufacture residential telephones in Singapore.
- A joint venture with several companies in Taiwan to produce digital switches.
- Lucky-Goldstar in South Korea to produce switches and fiber optic equipment.
- Rockwell, Honeywell, and Data General for swapping data between computers and switches.
- Telefonica in Spain to design and manufacture integrated circuits.

We appear to be witnessing the first stages of growth of global business alliances that put together major corporations from the United States, Europe, and Japan into something that will be more than the joint ventures of the 1960s–70s. What differentiates these alliances from ordinary joint ventures is that the latter were rarely a bevy of integral pieces in a singular grand strategy, forging a new "industry" in a new economy. Traditional joint ventures, even the most successful, had very little effect on the product-, market-, or organizational-space of the originators. The new alliances, however, are responses to the need to do just that. Even the most giant of corporations are reconfiguring the basic dimensions – space – of their organizations to adapt to the necessities of the new economy.

Whether organizations shrink through down-sizing, grow through alliances, or remain the same size, they will nevertheless be reorganizing their internal space. When you divide a whole into parts, it is the space between the parts that unites them together. Space is intangible and, as we have seen, intangibility is increasingly prominent both in the new economy and in its new organizations. The industrial image of structure, for example, is the girder-like architecture of buildings. The image of structure in the new economy, however, will be more like the architecture of atoms, built on energy and information, not steel.

Industrial organization structures, like buildings, are a hierarchy of boxes and lines. 2001 organization structures, like atoms, are a network of relationships. Marilyn Ferguson expresses the difference quite well:

> The network is the institution of our time: an open system, a dissipative structure so richly coherent that it is in constant flux, poised for reordering, capable of endless transformation.
>
> This organic mode of social organization is more biologically adaptive, more efficient, and more "con-

*scious" than the hierarchical structures of modern civ-
ilization. The network is plastic, flexible. In effect,
each member is the center of the network.*

*Networks are cooperative, not competitive. They
are true grass roots: self-generating, self-organizing,
sometimes even self-destructing. They represent a process, a journey, not a frozen structure.*[2]

In the hierarchical model, for example, we are seeing
the shrinkage of middle management; in the holistic
model there is no middle to shrink. Pyramidal organizations talk of the demise of middle management, and
all middle managers naturally see this as a *danger*. In
organizations based around network, there is, concep-
tually, no hierarchical middle; therefore, there is the
opportunity to define a new role for "middle" managers.
Obviously, organizations are not going to eschew all
hierarchy. The potential and practical gain in shifting
perspective from hierarchy to network is in the redefining, therefore, of who is in the middle.

Specifically, this means redefining the middle away
from being a layer of employees in between two other
employee layers and toward being a link in between the
provider and the consumer. From this new perspective,
the middle in organizations that emphasize the con-
sumer-producer relationship are the people who directly
link the two parts (see Figure 2).

Figure 2

Who Is in the Middle?

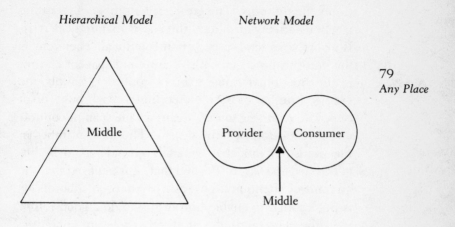

Hierarchical Model

Network Model

Middle

Provider

Consumer

Middle

In the old conception, middle managers are those below senior managers and above supervisors in a hierarchy. In a holistic conception, it is more effective to think of people in the middle as those who have direct responsibility for the user-provider relationship, regardless of where they are in the hierarchy. These people are truly in the middle, that is, in the middle of the customer-employee relationship in the business, rather than in the middle of the employee-employee hierarchy in the organization.

Another aspect of rethinking internal organization space is that corporations can shrink in their internal structure as well as in their total size, even in the midst of their own economic growth. Hierarchy is the backbone of organization in the industrial model and, while it will continue to occupy a major role, it will also shrink in both size and importance. The reasons this will happen are linked to the enablements of information technology, not to a shift in values.

A hierarchy is a classification according to rank, capacity, or authority. The greater the number of reporting levels in an organization, the more hierarchical it is. Bureaucracies tend to be very hierarchical. Hierarchy is the vertical dimension of structure and span of control is the horizontal one. Span of control is simply the average number of people reporting to each boss. A lot of people reporting to one means a wide span of control, and a few means a narrow span. The ideal number in the industrial model is generally considered to be between five and seven. Beyond that, it is believed that the amount of attention a boss can pay to each subordinate is not enough to enable him or her to do a good job.

In the classical model, if all other relevant characteristics of a business are held constant, the vertical and horizontal dimensions are inversely related. This means that when too many people are reporting to an individual, an intermediate level of supervision can be inserted in between; the span of control shrinks, and the hierarchy increases. Conversely, when the hierarchy shrinks, the span of control increases proportionately.

Many reporting levels shield those at the top from both petty annoyances and vital information. Too few reporting levels does the opposite; leaders can get inundated with trivial as well as important information. The military is very hierarchical. The Catholic Church is the classically flat structure, with only four levels: Pope, cardinals, bishops, and priests. Few, if any, business corporations are as flat as the Church, though more than a few are as hierarchical as the armed services. In the new economy, this proportion will gradually reverse. Let's see why.

In the industrial model, expertise generally increases as you climb the hierarchy. Expertise also resides in the specialists, the staff of the organization, and they are necessarily a small number of folks relative to the total employment of the organization. By transferring their expertise onto software, onto the expert system, anyone who has the expert system is the expert. As these expert systems get linked up into local and long distance networks, the generalist can access them immediately. While in customers' homes or offices, for example, salespersons can access many specialized databases and related "expert" services in order to close on the sales during the calls. They will simply connect their portable microcomputers with built-in modems, via the telephone, to the expert provider. This is the any time, any place expert.

Specialist staffs inside organizations, including swollen data processing units, may become significantly smaller for the same reason. General line managers can get the same staff work done instantaneously by external expert services at competitive market rates. These rates will almost always be below in-house rates, even where charge-back systems are used. Cost factors are also sending factory-like paperwork abroad. To avoid $9 an hour costs for keypunch operators, for example, American Airlines flies a daily average of over 1,100 pounds of documents to Barbados. There, for $2.20 an hour, the ticket information is transferred to magnetic tape, and the electronic data is then beamed by satellite to American's central computer in Tulsa.

Although the technological capabilities for creating tomorrow's organization already exist; nevertheless, I expect that it will take decades to actually mature because protectionist staffs will resist internal pressures for greater efficiency, just as protectionist tariffs resist external pressures. First the mindset has to change, to stop seeing a resource as a constraint.

Outside specialists, hired by corporations, face the same dilemmas and opportunities as internal staff experts. Many of their services will be available for purchase through expert systems. They face the potential shrinkage of demand if they resist and compete with these expert technologies of the future. From another context, however, these technologies offer them the tools to create new delivery systems for the services they can continue to control – if, once again, they view them as opportunities rather than threats.

Psychiatric screening is a good case in point. Most corporations that use screening subcontract the work to relatively small companies specializing in the task. These companies gained primitive economies of scale early on by advancing from face-to-face qualitative appraisal to paper-and-pencil type test scoring. Then the technology froze. Until recently, the mechanics have been applied to the scoring only. Now, interactive tests, using an expert system and network-connected microcomputers, can provide both probing richness and scale economies simultaneously – any time, any place. *Other* psychiatrists are developing these technologies for outpatient counseling of garden variety neurotics. Early prototypes sold to HMOs suggest that they provide more effective treatment than traditional first visits, and for one-quarter of the cost.

In each "industry" there is room for an aggressive leader to initiate organizational planning, using the any time, any place methods made possible by the new technologies. For the leader, however, it takes a decade of lead time for the organizational change to evolve and mature, and followers take even longer to adopt the new organization forms. The reason for the lag is not costs. You cannot spend your way into a transformed organization. The reason for the long wait, before these new organizational characteristics become present, is because, in many companies, the technological enablements are only now coming on-stream in meaningful ways.

Until the products and services that develop from the new technologies get past their own early growth phase, there is little reason to expect that the transformations will filter even further downstream into organization. But filter down they will, and if you want to glimpse the shape of organizations in the future, look to the constraints of time, place, and mass that are eased by technological developments. These enablements will be evident first in the products and services of the business, and in the processes used to produce them. Only later will they become evident in organization. For this to happen, first we must realize that organization is an intellectual phenomenon independent of time, space, and mass.

Key to this intellectual shift, for example, is the dilemma that all forms of organization have two simultaneous needs that are often at odds with each other: freedom and order. Freedom springs from intuition and leads to innovation. Order stems from intelligence and provides efficiency. Both are essential, but are they compatible with each other? Within corporations, these requirements are translated into structural terms with which we are rather familiar. Freedom is translated as the specialized interests of different parts, the optimal goal of decentralization. Order is represented as the regulation and integration of all those parts in harmonious and common action, the optimal goal of centralization.

The problem with the centralization-decentralization debate, however, was that the more you realized the benefits of the one, the less you got the benefits of the other. The dilemma of organization in the industrial corporation was the dilemma of an either-or world. Specialization, for example, had to be defined as either by function, or by product, or by area, or by market. Corporations had to select one of these dimensions as primary, and then subdivide the other three in turn, if necessary, into subordinate units further down the pyramid. The appropriate choice of primary, secondary, and tertiary dimensions is largely the result of strategic needs in each case, but in all cases it is also the result of an implicit assumption that the entire corporation cannot specialize by two or three dimensions *simultaneously*.

Will organization models evolve in the new economy that allow both decentralization and centralization, specialization and integration, simultaneously? The answer to this question is probably yes. It will take a very long time, but the enabling technology is already present. What is holding us back is our inability to see organizations structured spatially, around two or more dimensions at the same time.

The ability to see opposites existing simultaneously began in science around the turn of the last century, when there was a tremendous debate in physics about the nature of light. Einstein, using the photoelectric effect, "proved" that light was composed of particles. But a century before that, Robert Young used interference patterns to "prove" that light was made up of waves. Each experiment was valid and true, yet each contradicted the other, without being able to disprove or reject it.

The wave-particle duality was a very untidy dilemma, which forced physicists into abandoning either-or ways of looking at physical reality and turning to the probabilistic world of quantum mechanics and relativity theory. These scientific discoveries about the nature of the universe supplanted Newtonian mechanics with the technologies of the Einsteinian age. The products that were built from the new technologies came into existence only because the new context of physical reality had been accepted. This context accepted the simultaneous existence of mutually contradictory phenomena, without trying to resolve the contradiction.

Western management has not been very successful at building such simultaneities into its models. This is hardly surprising, since Western science still has a tough time of it. And while management continues to live in the either-or mechanistic world, it will continue to view organization as full of irreconcilable trade-offs.

There is a story about a king who had no heirs, and was fond of food and intellect. He offered the inheritance of his kingdom to whoever could create the best dish of food that was both hot and cold at the same time. Most people were stumped by the apparent contradiction. The winning dish was the hot fudge sundae, the runner-up created Baked Alaska. Managers who can hold opposites in their vision simultaneously can win the kingdom.

This contextual shift is exactly what management has to undergo before it can move into the new organization models of the new economy. Alfred Sloan's model in the 1920s, centralized planning and financial control simultaneous with decentralized operations around a product division structure, was an early step toward a pluralistic model. A more recent attempt, though badly flawed, has been the matrix.[3]

Matrix is a method of organization that rejects the age-old precept of "one man – one boss" in favor of a multiple command structure; that is, two or more bosses share power over a common subordinate. The idea behind the form is that it is workable to organize along two or more dimensions (functions, products, areas, markets) simultaneously, and that the benefits of each are not mutually exclusive. Borrowing from political models, it attempted to replace a single hierarchy of power and a unity of command with a balance of power. Matrix was popular for about a decade, the 1970s, but never truly lived up to its promise. It is more than a fad, however, and is still used in limited fashion in many corporations and widely in project-driven businesses.

No other organizational form has yet evolved to replace the singular industrial hierarchy with one that allows for contradictions. The best candidate to do so in the future is networking. Networking entered into the world of those who study organizations as a reaction to the formalism of classical theory. The classical theorists emphasized structure, and focused heavily on hierarchy. This meant understanding the either-or trade-offs between manageable spans of control laterally, and the number of reporting levels vertically.

The reaction against this aseptic approach was to stress human relations and the informal organization. In this context, networking meant the informal web of relations people developed with each other inside the organization. If you want to get anything done inside a bureaucracy, for example, you need a network of informal contacts rather than the book of formal rules. In other words, the term is not about informal relations so much as it is about how to put different parts of the organization in touch with each other so as to get things done.

In the age of information technology, the term networking has come into usage again. It still means the same thing, different parts of an organization in touch with each other to get done the job of the whole. This time around, however, the network is very different because it relies, not on an informal web of personal contacts, but on a technological web of information-handling systems. The way in which information systems are constructed enables us to avoid the either-or dilemma of hierarchy and to construct wholes where all parts are equally and simultaneously accessible to one another.

Both the social network and the information network provide the glue that holds unconnected parts together. But, where the social network operates to overcome constraints imposed by the spatial arrangements of hierarchy, the information network sees no such constraints, from either distance or hierarchy. Any parts of the whole that need to communicate with each other do so instantaneously. Networking means hooking together people who are geographically and hierarchically separated so that they may communicate with each other quickly and directly. The technological developments of computer networking have focused on communication across geographical space. The same developments will also enable communication across organizational space.

In the old hierarchies, for information to flow between people of different levels in different chains of command, it would have to flow up to a common boss and then all the way back down again; or, at best, it would have to travel to the switch point where the higher level traveling laterally and the lower level traveling vertically connected. For a salesperson to communicate customer information to the manufacturing plant, for example, the communication would flow up the sales chain of command to the general manager and then down through manufacturing until it reached the intended party – usually too late, and garbled. At best, some earlier level in sales might have directed it laterally over to manufacturing. This formal design is well known, as are the informal methods designed to overcome the limitations.

Information technology provides a formal method for overcoming the limitations. Organization structures and systems both deal with the same issue: how to create an architecture that best relates together the many parts of the whole. The early architecture of both was hierarchical, with the limitations we have discussed. Systems architecture has advanced from hierarchical principles to ones based on networks and relations. Structural architecture has not advanced this far. The simultaneous integration of independent parts, therefore, is served better by the systems than by the structure. Where the structural matrix based on humans failed, the electronic matrix based on information technology may yet succeed.

Databases were originally designed hierarchically. If you wanted to create a database of students, classes, and classrooms in a school, for example, you would have to predetermine which of these three dimensions to sort them into first, which second, and which last. The result would look like a hierarchical organization chart. If you had sorted first by subject, then by classroom, and finally by child, and you wanted to locate Johnny, you might have to trace down a number of dead ends until you found him.

With the new information architecture, you could find Johnny much more quickly, because information stored using semiconductor memories can be read in any order. The time required to obtain the data is independent of the location. In the previous example, so long as classes, classrooms, and students are specified as the dimensions for dividing up the whole, then Johnny can be located instantaneously because he belongs to all three dimensions simultaneously.

Electronic information systems enable parts of the whole organization to communicate directly with each other, where the hierarchy wouldn't otherwise permit it. What the hierarchy proscribes, the network facilitates: each part in simultaneous contact with all other parts and with the company as a whole. The organization can be centralized and decentralized simultaneously: the decentralizing mechanism in the structure, and the coordinating mechanism in the systems.

Networks will not replace or supplement hierarchies; rather, the two will be encompassed within a broader conception that embraces both. We are still a long way from figuring out the appropriate and encompassing organization models for the economy we are now in. At the very least, it is clear that we will have to reconceptualize space, transforming it by technology from an impediment to an asset.

No-Matter

"What's the matter?"

"Never mind."

"What's mind?"

"No-matter."

E $=mc^2$. If the equation governs the universe, and management is part of the universe, then what does the equation tell us about management? If we transpose the equation to $m = E/c^2$, it tells us that matter is nothing more than energy slowed down to a velocity that the mind can comprehend. Therefore, if you are concerned with the new economic reality, pay primary attention to the intangibility of matter; and if you are concerned with the new managerial reality, pay primary attention to the intangible attributes of your business and its organization.

Matter is not all that matters! In the new economy, the value added will come increasingly from intangibles, "things," whose importance does not lie in their material existence. As marketing maven Theodore Levitt puts it, "Everybody sells intangibles in the marketplace, no matter what is produced in the factory."[1] The service economy is dematerializing what does matter.

Take money, for example. Money is a medium of exchange used to establish comparative values in the market. Different economies select different commodities as the basis for this medium. Precious metals were used in preindustrial economies, paper money was basic in the industrial economy, and in the postindustrial economy we have turned increasingly to an intangible medium of exchange, as in EFTS (electronic funds transfer system). MasterCard, today, is putting a 64K RAM chip on your credit card, so that you can fill it up electronically with money and all other sorts of information. Instead of having to write a check and fill out numerous claims forms after a visit to a doctor's office, you can hand the receptionist the card, which will process all the information and money electronically.

Shaving in an agrarian economy was done with a straightedge razor, the safety razor of the industrial economy was invented by King Gillette, but the disposable razor is the no-matter creation of the new economy. In the photography business, film and cameras seem to be competing with each other for which one will make no-matter out of the other the fastest.

Like the disposable razor, Fuji came out with a disposable camera. For under $7, the consumer buys a roll of 35mm ASA 400 film, housed inside a throw-away camera of top quality. They've sold millions of them in Japan, and have entered the U.S. market this year. Kodak apparently had the technology but, deeply steeped as it is in industrial paradigms, didn't see a market for the no-matter product until now. Viewing no-matter as a resource allowed Fuji to be proactive; and not seeing it that way kept Kodak in a reactive mode.

Canon, another Japanese company, is going about the no-matter trend in reverse: They were the first to market a fully electronic camera that snaps color pictures without film. Right now, the electronic camera is not cost competitive, and the image is still too granular; but it could, in time, displace film. Either way, photography in the new economy is headed toward more no-matter.

The mundane hotel room key offers yet another example of the product shift to no-matter. The room key of the industrial economy was metal, and that metal was the only value in it. The hotel room key of the new economy, a small piece of cardboard or plastic, has less mass, but a great deal more value than its more tangible ancestor: Besides the key being lighter and cheaper, the room lock can be changed instantaneously for each new guest, ensuring much greater protection.

Intangibles are a part of every business, and increasingly so in the new economy. But until we are more aware of the full significance of this, we will continue to see intangibles as just that: *In* the economy, but not *of* it. It is only in the past decade, it seems, that we have begun to grasp this distinction. It is still the professor of marketing more than the average manager who stresses intangibles. Between now and 2001, however, the importance of no-matter will become an intrinsic part of our thinking and planning. Managers who consciously work with the no-matter of their businesses will begin to pull ahead of their matter-minded competitors, both within the organization and in the marketplace. Those who formulate and implement businesses only for tangibles will be working with less and less of the value in the economy.

Both products and services can be intangible. Because services now dominate the economy, I will emphasize these. Nevertheless, we need to be aware of the intangible aspects of products and of the need to make tangible the innate no-matter of services. For both products and services, however, knowing whether the intangibles are at the core or at the periphery of your business deliverables is central to developing a strategy in the new economy.

Businesses in the new context are different from the older industrial-based ones, to the degree that they allocate increasingly intangible resources to provide incrementally intangible products and services, for ever more intangibly defined markets. Customers do not need tangible goods less, they need intangible goods and services more. In addition to the absolute growth of intangibles, growth in the new economy means an increase in the value added by intangibles relative to tangibles. I am not leading a cheer for the fact; it is simply so, and we have got to incorporate it into our management and organization – which are also increasingly intangible.

We are familiar with allocating tangible resources (labor, capital, equipment) to create tangible products (food, cars, houses) and tangible services (restaurants, telephones, cleansers) for tangible market segments (middle class male professionals, teenage suburban athletes).

By comparison, we are less practiced in allocating intangible resources (mind, time, information) to create intangible products (software, advertisements, investments) and intangible services (personal shoppers, health, education) for intangible market segments (conservatives, compulsives, swingers).

As things stand now, we are not very good at dealing with intangibles in business, even worse, we don't seem able to grasp what intangible organization might mean. We do not have adequate managerial concepts and theory to do so. If we want to develop these tools, Alfred D. Chandler, Jr.'s Pulitzer Prize winning book, *The Visible Hand*, about the managerial revolution in American business, is a good place to start. Chandler shows how industrial organizations took the place of market mechanisms in allocating resources and coordinating the activities of the industrial economy. "In many sectors of the economy the visible hand of management replaced what Adam Smith referred to as the invisible hand of market forces."[2]

Chandler saw, as we do here, managerial and organizational forms as following the transformations of the economy. During the early industrial era the emphasis was on the external forces of the market; this shifted during the later industrial era to a focus on the internal forces of management. For Chandler, the early external market was an invisible force on business; during the final decades of the era, the force on business, internal management, was visible. In other words, forces become visible only as an economy matures. In the early decades they are far more elusive, abstract, and unsubstantial. We are still in the early decades of the new economy, so we are experiencing the shift from the tangible and visible hand in industrial organization to the intangible and invisible hand in the newly forming economy. By 2001, when the new economy has matured, we can expect that the new intangibility itself will be more visible.

One step in this direction would be for us all to gain a more realistic grasp of this shift, away from the production, distribution, and consumption of tangible goods toward intangible services. So far, the major classification scheme is the SIC code, created by the Bureau of Economic Analysis of the U.S. Department of Commerce. Used particularly to classify employment and GNP, SIC stands for "Standard Industrial Classification," which we can see from Table 1. It is now a misnomer, insofar as the majority of jobs and output are no longer from "industrial" production.

If you are a lawyer, for example, working for General Motors, you are counted as part of the industrial sector, because G.M. is considered a company in industrial manufacture. If you leave, move across the street, and hang up your own shingle, doing the exact same kind of work, and only for G.M. (in other words, G.M. is your customer, now buying rather than "making" your services), you are considered to be working in the service sector.

In other words, all staff support and service personnel in what we call "industrial" firms are counted as part of the industrial economic sector. Clearly, this is very misleading. Thomas Stanback and others estimate that at least 5 percent of the GNP in the U. S. is created within

Table 1

U. S. Employment and GNP, by Selected Years, for
Agricultural, Industrial, and Service Sectors

Year	Sector	Employment	GNP
1865	Agriculture	48%	22%
	Industry	14	22
	Service	38	56
1929	Agriculture	8	9
	Industry	37	28
	Service	55	63
1945	Agriculture	4	9
	Industry	39	35
	Service	57	56
1985	Agriculture	2	2
	Industry	21	28
	Service	77	70
2001	Agriculture	2	3
	Industry	5	24
	Service	93	73

the central administrative offices of firms. Their output
is in the form of intermediate producer services, that is,
services that add value to the offering yet are consumed
within the producing firm rather than by the ultimate
customer.[3]

Only 6 percent of IBM's employees worldwide are in
industrial manufacturing. Less than 20,000 of its
400,000 people worldwide are production employees,
yet IBM is nevertheless (mis)classified as being part of
the industrial sector of the economy.

Three things are happening simultaneously. One,
there is an overall employment decline in manufacturing

firms. Two, servicelike employment is increasing *within* manufacturing. Three, it makes less and less sense to dichotomize between industrial and service companies. The grand old names of industry are less and less industrial. A third of the revenues at General Electric and Borg-Warner, and half of those at Westinghouse, come from services.

There are many misconceptions and problems in measuring the output of services. One particular problem involves recognizing the joint nature of an offering. There is often a significant service content in the final output of goods, that is not accounted for because traditional national income accounting procedures are biased in favor of tangible goods. According to Michael Packer, industrial techniques stressed the "efficiency of outputs"; the new economy needs to focus on the "effectiveness of outcomes."[4]

Despite the antiquated classification system, which *under*counts service work, basically three-quarters of our economy still falls into that category. If the national accounting frameworks were more accurate, the proportion would be even larger. Using the same reactionary method, the division in 2001 becomes skewed even to the point of absurdity. Economic classification schemes have only very limited utility when one category encompasses between 75 and 95% percent of the cases.

A contextual shift is needed. As we saw at the beginning of the book, agrarian, industrial, and service economies each have agrarian, industrial and service sectors. The most important perceptual transformation we can hope for is for everyone to realize that *services play a major role in every sector of the new economy, and not just in the service sector*. Industrial manufacture played a main role in mechanized farming; in the same way, services in the new economy are a large part of the industrial sector. Advice about what product to buy, instruction about how to use it, and repair when something goes wrong are all service elements that represent a significantly larger part of the total offering in today's economy than they did during the industrial era. A study prepared for the Association of Field Service Managers, for example, estimates that a well-run service operation can contribute as much as 30 percent to a manufacturer's revenues.[5]

After Adam Smith, the next giant economic theorist was Karl Marx. What would he have thought about all this? According to Marx, ownership of the means of *material* production was what counted. Production of the immaterial, the production of intangibles, was synonymous with unimportant. Postindustrial capitalism increasingly will be nonmaterialistic, not as Marx expected to see it, but in the sense of increasingly valuing intangible "things." The next great theorist, first of the postindustrial period, was John Maynard Keynes. His theories on deficit spending for prosperity, that is, spending what does not yet exist, certainly moved us further into the intangible economy. The next great theory will have to account for still another major transformation, that is, in the new economy, both inputs (resources) and outputs (goods and services) are increasingly intangible.

To businesses, resources are generally thought of as assets that are used directly or indirectly to create products and services. In the industrial economy, resources are largely tangible matter; however, in the new economy they are increasingly intangible; nonmaterial, though hardly immaterial.

Typically, in any economy, matter is seen two ways: according to its physical properties and according to its economic ones. From one perspective, matter is substance; from the other, matter, although it is substantial, is viewed as either important or inconsequential.

In the industrial model, what is worthwhile is generally tangible; the two adjectives seem to go together. Our speech reflects this when we say, "a man of substance" and "a solid citizen." Despite this nod to the value of individuals, when it comes to assessing their productive worth, our industrial frameworks are woefully inadequate.

It is worth stressing again: We need to distinguish among the sales, assets, and earnings contributions of intangible *human* resources just as the industrial models do for tangible *physical* resources. In the new economy, tangible products and services rapidly lose their value, and the larger value added comes from intangible products and services. Early computers, for example, were very valuable products. Now they are rapidly becoming commodities. The value added is in the intangible software, in the solution and the service. Computers will become like calculators and pencils, where the tangible product is the cheap and simple means used to accomplish the more valuable and intangible end.

An important piece of the intangible value added comes from the asset value of knowledge residing in the employees' skill base. When capital is allocated to a business, an increasing portion of it will be invested in intangibles such as the knowledge worker.

The asset value of knowledge may also reside in the skill base of the customers, when that can be tapped. CompuServe, for example, is a firm that makes available around 1,500 different databases to its almost quarter-million subscribers. CompuServe invests in intangible assets and lets the customer make the investment in fixed assets, in the form of the personal computer. In so doing they have transformed the tangibility of their assets in two ways: one, by shifting tangible assets into intangible information; the other, by moving the fixed asset investment into the hands of the consumer. (This is parallel to the shift in locus of manufacturing, discussed in the chapter Any Place.) Business systems based on new technologies are making it possible for companies to reduce further their fixed assets. "Throughout the world," says James Heskett, "manufacturing is substituting information for assets. Nearly every program to reduce inventories has this character."[6]

Also, since it is appropriate to account for the depreciation and obsolescence of physical (tangible) assets, why not do the same for human (intangible) assets? In declining businesses, intangible assets decline just as much as tangible ones. The adjustment costs of corporate and economic transformation pertain to retraining and relocating as well as to depreciation schedules and plant shutdowns.

Resources are generally finite, but information is infinite, and the greatest value-added in today's new economy. Whereas hydrocarbons were the major fuel for the industrial economy, information is the major fuel that provides energy for the new economy. Furthermore, knowledge is the principal product produced by this fuel. Although there are limits to growth in finite resources, there are no limits to learning, the ultimate renewable resource.

The more information you add to a finite resource, therefore, the more valuable it becomes. The economist Paul Hawken goes so far as to make information and resources synonymous. According to Hawken, the reinforcing ability of concrete and the atmosphere of the restaurant are both information. Price, quality, design, utility, and workmanship are all pieces of information about products and services. "The single most important trend to understand is the changing ratio between mass and information in goods and services."[7]

In this very succinct way, Hawken captures the essence

of the shift from the industrial-based mass economy to the information-based service economy. Expanding Hawken's ratio to an equation, we can say:

$$\text{VALUE OF A DELIVERABLE} = \frac{\text{INFORMATION}}{\text{MASS}}$$

Steel has a lot of mass with relatively little information, contrasted with a computer chip which has a lot of information relative to its mass. If a sheet of paper is coated, it has more information, in Hawken's sense, and is therefore more valuable.

My award for the most creative dematerialization of matter goes to MCI for their invention of the "electronic ounce." Within classes of service distinguished by speed, the U.S. Post Office charges for its services by weight. It costs 22 cents to mail the standard one-ounce letter. MCI is competing with the postal service by advertising that it can send the same letter electronically, faster and for less money. Its electronic ounce is the amount of intangible information transmitted for an equivalent price.

Most information is generally consumed as an intermediary good. Except for the information sector, whose purpose is to produce communicable knowledge as an end in itself, most information is used as a tool, support, or medium in creating final goods and services. However, as the intangible proportion of the value of goods and services increases, information too will become a more useful end-"product." The humane effect on society, of information being used more as a final good than as an intermediary good, is worthy of serious attention in the future.

For our purposes here, a product is an idea in a tangible form, and a service is an idea in an intangible form. As tangible objects, products exist in both time and space; services exist in time only.[8] Because of their intangibility, services cannot be stored and carried in inventory. They must therefore be produced in real-time, at the instant of delivery. Production, delivery, and consumption are virtually simultaneous. From the customer's perspective, the service doesn't exist until it is delivered and consumed. For example, from the chef's viewpoint, a meal exists at the time he cooks it (production); from the customers' viewpoint, it exists when they eat it (delivery and consumption).

This instantaneousness means that quality assurance must go in before production, not afterward, as with manufactured products. This also makes it very difficult to demonstrate service by providing samples. One customer can see another customer getting good service, but the experience is indirect. You can test-drive a car without owning it, but it is very difficult to test-drive a doctor or telephone company the same way, and there is no money-back guarantee on bankers or real estate agents. Services have to be experienced directly; yet, because of their intangibility, they cannot be tested the way products can.

Despite this difficulty, however, producers can pretest a service more easily than a consumer can. Market research, for example, will tell the provider something about the conditions for acceptability of variable-rate mortgages; but buyers cannot pretest the mortgage the way they can turn on a television before they buy it. The intangible mortgage makes possible the tangible house, but only at the formal closing ceremony. Intangibles have to be experienced directly. That is not possible until their actual delivery, and once delivered they are essentially consumed. In other words, in the same way that they cannot be stockpiled prior to delivery, they cannot be sold as "used" afterward.

Another important distinction is that intangible services are performed in addition to being produced. They are essentially social actions, conducted between the deliverer and the consumer. Because the user participates in the creation and quality of the service, the deliverable has an intrinsically experiential quality to it. The experiencing of the service is part of the service itself. If the provider delivers a service that is negatively experienced, the service itself is negative.

For people in service occupations, therefore, an important distinction is whether they work hard to get the job done or work hard to meet the customers' needs. From a service point of view, the two are the same; the job isn't done unless the customers' needs are met, and when they are met the job is done. Unfortunately, the two are often different, and represent the difference between people who are and are not service oriented.

How many times have we each asked a waiter or a stewardess for a cup of coffee only to be told, directly or indirectly, that we have to wait until they do X, complete Y, or get through with Z first? When this happens, one often gets the feeling that these people could do a great job if only the customers wouldn't get in their way. They take their cue about what is proper from the internal system, rather than from the external source – the customer.

As services come to dominate the new economy, we will appreciate more and more that their essential characteristics are not limited to services only. There is a well-known story about the Stanley Tool Company, where the consultant – working with the orientation that we are suggesting – tells them, "You are not in the business of selling drills. You are in the business of selling holes." The tangible product is used only to fulfill the intangible need. With the focus on the desired ends, the means are not fixed; they vary with the latest technology. Early drills were mechanical, then electromechanical. What is the next likely technology for making the best holes? Probably lasers. It is not unlikely that by 2001 we will be buying laser drills, and maybe not in the old "hard"ware stores but in new "soft"ware stores. The laser is composed of energy that is not slowed down by the square of the speed of light. The result is a powerful, dematerialized drill.

Olivetti is well known for the aesthetics of its design. In the past, design meant design of the box, the hardware. Since customers buy the hardware not to have pretty boxes but to run their software, Olivetti's challenge will be to build aesthetics into the design of the software. "Not because it is pretty," as one of their executives said, "but because it is powerful." Tomorrow, the pleasantness of interacting with the box will depend on the design of the software. The future of design lies in designing the functionality and the aesthetics of the intangible. Olivetti is also very interested in ergonomics, yet this relatively new field has thus far been an ergonomics of hardware. If it survives, the future of ergonomics, like that of design, will be in the software.

Furthermore, whether they are made of matter or of no-matter, quality products that are negatively experienced will have negative after-sales results. The experience of a manufactured product is real and important for both tangible goods and intangible services.[9]

Because products are "things," whereas services are acts and interactions that must be participated in and experienced to become real, it is useful to think of the production-consumption of a service as a social event. Emotions, values, perceptions, attitudes, expectations, and other human traits of both the customer and the employee come into play more than they do with products. And there are more opportunities not only to manage, but to mismanage the exchange. Quality and value are largely subjective.

One simple rule of thumb is, the fewer people there are between the provider and the user, the greater the likelihood of user satisfaction. This is because the buyer not only gets and consumes the service but also participates in its production and delivery. When the exchange is direct, the distinction between *producer* and *consumer* breaks down. This is what Alvin Toffler meant when he coined the word "prosumer" to describe the integration of the two actors.[10]

For the customer to take on part of the producer's role is a positive and innovative characteristic of the new economy. However, in large organizations there is an opposite tendency to interpret this negatively. It is a holdover from an earlier, industrial period, a misconception that the line is the staff's customer.

In fact, *there is only one customer*, and that is the customer in the marketplace, whether it be a corporation or a consumer. Only a small percent of a company's employees may have direct contact with customers, so many of them come to think of other employees as their customers. This is typical in staff positions, where an ethic often develops that they are "there to serve the line." I have sat through discussions in dozens of companies where people speak of "internal customers." It may seem like a wonderful way to spread a service orientation throughout an organization, even when there are no bona fide customers, but I believe that the opposite is true. Employees who think of other employees as their customers lose sight of the business in favor of the organization.

In large corporations, there are often four or five links between employees before reaching an employee with direct customer relations. The notion of internal customers, therefore, is bound to get watered down in the process, to the point where actions taken may have zero or negative relevance to actual customer value. Even where there is some value, just how much will never be known. It will in fact be so buried in the affairs of organization that its relevance to customer needs will be impossible to assess.

In some service-oriented companies the message is, "If you're not serving the customer, you'd better be serving someone who is." This is helpful because it reduces the number of organization layers and puts the action where it ought to be – on service. But the focus is still not on the customer.

For employees who have no direct customer contact, a more healthy service attitude is to approach those who do with the orientation, "How can I assist you in serving our (real) customers?" That way, every employee, no matter how many links removed from the customer, is truly service oriented. Can you imagine how much more powerful a company would be if every employee, no matter what their job, was focused on serving the customer in the marketplace? if each employee, no matter how removed from the frontline, asked, "How does my job fulfill our customers' needs in the market?"

The industrial model focused on the producer. The early postindustrial model, a metamodel for the new economy, tells us to focus on the consumer. The business model for the new economy, however, should increasingly focus on the relationship between the producer and the consumer as the most relevant space. Excellent companies, an overused term used for the first time in this book, are not those that put the customer first or the employee first. With that orientation, whichever comes first – the other doesn't. Instead, excellent companies use a holistic approach, that is, they put the customer-employee relationship first.

The contradictions that are so prevalent in mechanistic models, like emphasizing either the customer or the employee, do not appear in holistic models because in the latter opposites do not belong to different wholes but are treated as extremes of a single whole. All phenomena are manifestations of a continuous oscillation between seemingly opposite parts.[11] From this perspective, tensions and distinctions between producers and consumers, for example, are encompassed in a larger whole. The focus is on the nature of the relationship between the two.

This is not an orientation for service businesses as opposed to manufacturing concerns. In the context of the new economy, it is, for service and manufacturing, simply good business. In the same way that service businesses were managed and organized around manufacturing models during the industrial economy, we can expect that manufacturing businesses will be managed and organized around service models in this new economy. Context creates a reality, and the context created by the new economy is service. Context is intangible, and service businesses fill intangible needs with intangible products. This new economic context was there for three or four decades before we discovered it.

Now that we know that we didn't know about it, we are making up for lost time, rapidly focusing on the preponderance of service businesses in our economy. We are also beginning to take an active interest in the management and organization of these businesses, seeing them as needing a new form and a new approach. The industrial forms were an improvement over the preindustrial models for running economic enterprise. In the same way, utilizing new knowledge and new technology, the new forms of organization we create will be better.

The forms will speak to what we have discovered is the largest sector of the new economy—services. Certainly, one of the new frameworks will ultimately redefine the economic sectors, so that services is no longer a category in and of itself but rather a pervasive element of all categories. For reasons like these, the new managerial and organizational context will start to dominate all businesses, those that produce product matter as well as those that service no-matter.

From the science of the universe through technology to business, organization is the youngest of the lot; the last to develop. One of the realizations we need to come to about organization is that we do not just shift from one context to the next, from industrial models to service models, but we let "context" be the context itself. In a mature economy, managers manage content within frameworks they take so much for granted that they don't even question their boundaries or appropriateness. In consequence, the frameworks must become terribly inappropriate before people will begin to say, "Hey, we need new models." Since Daniel Bell wrote *The Coming of Post-Industrial Society* in 1973 [12], we have been attempting to define the meaning of "postindustrial" in its own terms. We must do so for the economy and businesses, however, before we can expect to do so for management and organization.

By 2001 the new economy will probably be well into middle age, mature enough to understand itself. By the time it reaches its golden years, someone (or ones), as Sloan did in the industrial 1920s, will have come along and reset the context. By that time, managers will again be managing content within another fixed and unexamined framework. Until then, however, we are in a marvelous period of openness. We should not rush to close this lack of definition prematurely. For one thing, we can't, even though we may want to. The only thing we could do, but should not just yet, is label it and lock it in. It is much better to let it hang out on its existential edge for a while; to play with it, be flexible, innovate, and create.

Isn't that what all the executives and commentators

have been crying for? Manage the context; let your subordinates manage the content. And let them do the same; and so on down the hierarchy, until the technology of the new economy and the values of the new society together evolve into the network that is now replacing the old order.

From the perspective of business, management must ask, How does the service context meet customer needs? What are the key elements to be provided? And how are these elements perceived by all the relevant stakeholders, particularly the customers and employees? How do competitors define the market needs? And how does their service context differ?

From the organization perspective, the service context guides how the content of the service will be designed, marketed, and delivered. It will decide which elements of the service are core and which peripheral, which are explicit and which implicit, and which are tangible and which intangible.

The service context will tell you what business you are in and how to organize to manage that business. This is particularly important in industrial companies with natural rates of growth and decline, and with intentional programs of repositioning through divestiture and acquisition. These companies are increasing their service businesses faster than they are growing their traditional sectors. An example of this shift in business mix is evident in the core industrial business, autos.

In autos, the shift leads one to ask, when is a car company not a car company? As we saw earlier, the answer might be, When it is a bank. The three major automakers have been making car loans for a very long time. Not so long ago they started raising their own capital by issuing commercial paper. Now they lend money to other corporations, issue mortgages, and finance everything for the consumer from tangible products such as washing machines to intangible ones such as home mortgages and vacations. All these activities take the place of financial intermediaries. In fact, the carmakers have created their own financial subsidiaries, and rather large ones at that.

General Motors Acceptance Corporation, for example, is now one of the country's largest financial institutions, with $74.5 billion in assets – virtually as large as American Express, Metropolitan Life, or Manufacturers Hanover. Much of this growth comes from acquisition. Through this route, for example, G.M. became the nation's second-largest mortgage lender in 1984, with $22 billion in commercial and home loans on its books. In mid-1986, it suggested to many of its car and truck loan customers that they should also apply for home mortgages. GMAC's contribution to G.M.'s balance sheet is even more impressive: some 25 percent of G.M.'s $4 billion net earnings in 1985 and around one-third of its 1986 profits.

The picture is the same for Ford and Chrysler. Ford Motor Credit Co. has over $31 billion in assets, and Chrysler Financial Corp. has over $16 billion. Ford's credit arm represented 17.5 percent of earnings in 1985, and around one-quarter of its 1986 profits, while Chrysler's financing profits have about doubled, starting from a lower 9 percent base during the same period.

There are, however, many who argue that this sort of shift is not a natural evolution from one economy to another but rather a sign of profound sickness. Foremost among these commentators is Robert B. Reich, Harvard Professor of Government and author of *The Next American Frontier*. He says that the changes that have taken place have not been due to meaningful technological or institutional advances but are based on accounting, tax avoidance, financial shuffling, litigation, and mergers and acquisitions. This does not create new wealth, says Reich, "it merely rearranges industrial assets. And it has hastened our collective decline."[13]

Do these shifts create new wealth for the economy or only for the few wizards who play the mega-merger game? Clearly, the merger record a decade after the deals are done has not been outstanding. Also, it remains to be seen if General Motors, Sears, or American Express can provide better banking services than can standard commercial banks. But that is an organizational uncertainty; more a question of whether it can be done efficiently than of whether there is more value added into the economy as a result. As we said in the previous chapter, disintermediation can be a healthy curb to an antiquated economy. By 2001 we will know if the Sears et al. of the world have made our lives any better by one-stop shopping for both tangible and intangible services.

Another source of the growth of intangible services are those that contribute to long-term customer loyalty, even though they do not generate direct revenues themselves. An example of this is the use of 800 telephone numbers. Procter & Gamble, for example, answers more than one quarter of a million complaints annually. According to Karl Albrecht and Ron Zemke,

If only half of those complaints are about a product with a 30-cent margin, and only 85 percent are handled to the customer's satisfaction, the benefit to the company in the year . . . could exceed half a million dollars. Such a sum represents a return on investment (ROI) of almost 20 percent.[14]

According to the authors, the same is true for calculations on customer loyalty:

If automobile industry studies are correct that a brand-loyal customer represents a lifetime average revenue of at least $140,000, then the image of a manufacturer or dealer in a bitter dispute with a customer over an $80 repair bill or a $40 replacement part is plainly ludicrous. Similar logic holds for almost every business sector. . . . Appliance manufacturers figure brand loyalty is worth $2,800 over a 20-year period. Your local supermarket is counting on you for $4,400 this year and $22,000 for the five years you live in the same neighborhood.[15]

There is, then, in the economy an increase of no-matter from new service business growth in general, from the growth of service sectors within once predominantly industrial firms, and from services focused around the value of long-term customer satisfaction. Another way in which no-matter is increasing its slice of the economy is through recognizing and promoting the growth of intangible value within industrial products themselves. Here, at the micro-level, services become an ever-larger portion of the total product-service offering to the customer. Hawken's increasing ratio of information/mass is a case in point. More generally, this is the increasing shift in the ratio of intangible/tangible elements (i/t) in an offering to customers. The degree to which a business can increase its products' i/t ratios faster than the competition's is an important competitive advantage. Depending on the business, this may mean in absolute terms, relatively, or at the margin.

Any tangible product that cannot be experienced before it is purchased is bought on the producer's intangible promise and on the consumer's intangible trust. Even the money-back guarantee, a way of marrying the promise to the trust, is an intangible. Purveyors find tangible ways to get their customers as close to the experience as possible, but never the no-thing itself. The atomized spray of perfume on the wrist, for example, gives the shopper a sample of the still-tangible product but not the experience of attracting another person as a direct consequence. Charles Revson, founder of the Revlon Company, understood the difference between the tangibility of the product and the intangible experience of a fragrance. He made this cynically clear in his famous quip, "In the factory we make perfume, in the store we sell hope." Marketers have long stressed the importance of the intangible experience in such phrases as "Sell the sizzle, not the steak."

When Coca-Cola launched its New Coke, it was focusing on the tangible difference in taste from its near-century-old standard. What it totally missed was the intangible importance of the original formula to large portions of the American public. The reason Classic Coke was brought back was because it belonged to the American experience, as well as being a preferred taste.

How, then, do marketers understand the main elements involved in satisfying customers' needs? What many of them have done is to distinguish between elements that are either tangible or intangible, core or peripheral, and explicit or implicit. The intangible, peripheral, and implicit dimensions of a business concept all add to the no-matter core.

Tangibles can be described easily in terms such as size, color, and material. Some intangibles can be described clearly and directly, as in a perfect safety record and a zero-coupon bond; but the descriptions of no-matter are often more vague and indirect than are those of tangible goods and services. It is therefore common to tangibilize invisible services, as physical evidence that they have been performed. The checklist hanging from the rear-view mirror of the rental car, and the paper strip around the toilet seat in the hotel room, attest to the care that might otherwise not be noticed.

When the core is intangible, the tangibilized evidence may be clear, but it is nevertheless peripheral. Peripheral products, services, and evidence are useful when all competitors offer pretty much the same core. Once added to the core, however, they come to be expected by the customer and are very hard to discontinue. Then they take attention away from the important focus, and they can even mount up to an opposite effect. For example, adding peripherals can make it impossible to maintain a low-cost producer strategy. In this case, it is far better to focus on the core: for example, to seek out additional low-cost segments, to standardize the offering further, to automate, and to perform as much of the core as possible out of the customers' sight. Even in differentiated strategies, where attention to customized needs is most important, add-ons can defocus. Whenever possible, it is far better to enhance the core itself than to protect the concept by adding peripherals.

The third distinction, between explicit and implicit benefits, is familiar to everyone. An implicit benefit of a liberal arts curriculum is that students learn how to learn. An implicit benefit of a particular restaurant is to be seen (or not seen) by important others. An implicit benefit of preventive medicine is lower liability costs. Sometimes implicit benefits become explicit, as when people say of IBM "When you buy their products you buy the company" and "No one ever got fired by buying IBM."

Since the intangibility of things is so central to the new economy, it is logical that segmentation schemes for targeting a new-economy market also rely on intangible characteristics more than on tangible ones. In the industrial economy, market segmentation for both tangibles and intangibles tended to rely solely on demographic factors such as age, income, and family size. A wealthy, middle-aged female or a young man with a modest income are fairly concrete types to locate. A method for targeting a relevant population in the no-matter economy has absorbed this earlier one and included another: the new method uses demographics, but in combination with the more intangible psychographics.

Psychographics puts values, lifestyle, and personality at the root of what people need and will buy. It uses more subjective and intangible measures of attitudes, beliefs, and opinions. It sees the demand for quality rather than quantity, and for experience rather than things. "In the experience industry," according to Brad Edmondson, an editor of American Demographics magazine, "automobiles, jeans and beer are mood-altering substances. They are gateways to the experiences of driving, fashion and patriotism."

The approach was pioneered in the early 1970s by the market research firm Yankelovich, Skelly & White, who began assessing consumer habits on the basis of social trends rather than just population data. Another market research firm, SRI International, developed its now widely accepted VALS system in the late 1970s. VALS (for Values and Lifestyles) divides Americans into nine groups according to their primary motivation, and gives each a descriptive label, such as "Belonger," "Societally Conscious," and "I-Am-Me." A well-known example of the VALS psychographic approach came when the advertising agency Young & Rubicam identified typical Merrill Lynch customers as independent-minded private investors rather than as those who follow the herd. Merrill Lynch consequently changed its advertising campaign from a herd of bulls to a solitary bull, and its slogan from "Bullish on America" to "A Breed Apart."

Ray Ellison Homes, in San Antonio, uses VALS to design and sell houses. "Achiever" women prefer small and efficient kitchens that are easy to clean, although they don't object to cleaning big, luxurious bathrooms. "Belongers" are more family-centered, however, and like big kitchens.

Psychographers have left the drab labels, such as "middle-class suburban," to the industrial-age demographers. Instead, they create a bestiary of "Militant Mothers," "Self-Made Businessmen," "Money and Brains," "Whole Earthers," and "Comparison Shoppers." Despite the innovativeness of the labels, however, none has caught on to the same degree as the more demographic contribution to the new economy segmentation: the Yuppie.

Direct mail advertisers sometimes use a zip-code based system. This assumes that those with similar demographics have similar psychographics. From this, lifestyle profiles are developed to match up advertisers with consumers targeted by their intangibles. One section of town, for example, buys specialty wines, watches "60 Minutes," and enjoys sailing, but doesn't buy CB radios or AMC cars. Another drives Japanese cars, goes hiking and backpacking, and watches "Hill Street Blues," but doesn't chew tobacco or watch bowling on TV. Even if their targets are not truly that homogeneous, advertisers nevertheless use tangible criteria, such as location, to create segments around intangible traits.

Fingerhut, the largest and most profitable direct mail-order company in the United States has a strategy based on definitive characteristics of the new economy: low fixed capital, information-based technology, distribution expertise, and well-defined market segments. Using its extraordinary database, for example, it has found that people using a middle initial are a better credit risk than those who don't, and people filling out order forms in pencil are a worse risk than those who use a pen. These kinds of intangible criteria help Fingerhut target their segments with great precision.

Intangibles also play an important role in delivering to targeted markets, particularly when services are involved more than products. Services highlight the importance of the delivery system because intangible market needs are fulfilled by intangible deliverables. Since services dominate the new economy, since they are performed, not produced, and since they don't exist until they are performed, everything depends on executing the performance well. This means that, in implementing strategies, delivery systems play a much larger role in the new economy than they did during the industrial period.

All of us have taken out a loan, flown on an airplane, used electricity, and eaten in a restaurant. We consider these services fairly simple affairs. And they are simple no-matters, until we chance to witness the operating systems involved in producing and delivering them. The delivery system is a combination of matter and no-matter. Tangible matter, in the form of equipment and facilities, and intangible technology, in the form of information and procedures, all get together with the people who administer them in what is the equivalent of production and distribution systems in industrial manufacture.

Effective delivery systems should be designed by those who must manage them.[16] Initially, this means extraordinary detail. Because the service is essentially a process, tools from process engineering have been used. Time and motion studies chart the flow of activities, particularly those that are invisible to the customer such as check clearing and baggage handling. The use of PERT charts are borrowed from project programming to do things like cost versus time or value analysis. Lynn Shostack calls this "service blueprints," and uses it to do such things as spot weaknesses in the process, and to set, measure, and adjust standards.

While such detail can be useful initially, ultimately, the system must be simple and uncomplicated to those who carry it out. A favorite example is the cleanliness of well-known theme parks. At Disney locations, for example, minimum-wage teenagers, who probably don't clean their room as well, are cleaning up papers and spills nonstop. In fact, everyone on staff picks up trash whenever they spot it. This is the only business where I have seen managers do this, and as second nature. As you can imagine, the garbage piles up pretty quickly at these places; yet customers hardly see any. That is because there are systems of underground vacuum-powered tubes for garbage throughout the parks. There are similar, simple, yet thorough systems for parking cars, taking tickets, moving lines, serving food, tending gardens, first aid, and a host of other activities.

Disney has long been thought of as one of the most well-managed companies in the U.S. It is a wonder that many of the simple and effective delivery systems, pioneered at theme parks, haven't been copied more. As a customer, for example, how many times have you been stuck in a slow-moving line and gotten annoyed because the person in the next line came much later than you, yet got served much quicker? Didn't you ever wish you had gotten on the faster line? It sounds like an Andy Rooney question: "Why do you suppose they don't do something about that?"

The simple solution, used at theme parks on a massive scale, is a single line, which snakes back and forth to use no more space, and which moves the person at the head of it to the available slot. When each customer has a shopping cart, the snake line would be too unwieldy. Zayres stores have acknowledged this service element;

they advertise that they will open up another cash register if there are more than three people in line. Yet, despite such available alternative service efficiencies, hardly a week goes by when we don't stand in at least one line of the aggravating kind.

Well-run service delivery systems give lots of attention to Andy Rooney-like questions about minutiae. Employees, for example, are crucial in service delivery systems, and they are examined minutely. Fifty years ago, a basic distinction among employees was whether they worked with their hands or with their minds. The distinction was a simple dichotomy; assuming that, since mental workers did not use their hands very much, physical laborers did not use their minds very much either. Today, we know that the work shift from industrial to service dominance is from the labor force to the mind force, or to what Peter Drucker calls the "knowledge worker." Whether people work in industry or in services, the large majority already work with their intangible personal assets, their minds and personalities.

Interpersonal skills, therefore, become an important job requirement, because so much of service depends on a social situation, on personal contact between the provider and the customer. Also, since customer contact is often with lower-level personnel, their people-handling skills may be more important than their job-related ones. Even when the technical skills take years to master, it is usually easier, quicker, and cheaper to teach a pleasant person procedures than to teach a procedural person to be pleasant.

Great care, therefore, should be taken in selection of the right kind of people for the service offered. We are not talking about "nice" or "friendly" people, but about positive personality attributes that match the service appropriately. "Charming," for example, may be an appropriate trait for a maitre d' but irrelevant for the bus boy;

"helpful" is more relevant to the waiter than to the chef, who should be "creative."

These attributes can also differentiate one company from its competitors. Consider three competing department stores, one of which promotes quality, another fashion, and a third, price. Each organization can then translate these values into its personnel practices. In hiring policies, for example, each company can look for its key trait in the way candidates dress. Further, the first company can prize low turnover and long tenure in its work force, while the second may accept turnover and prefer youthful employees; the third company can pay less than its competitors.

In other words, select only the kinds of people whose personal needs are met by fulfilling the service concept of the role or of the business. This means that personnel selection criteria should be precisely defined and explicitly related to the business definition. Let psychological intangibles fulfill service intangibles.

Because of the heavy emphasis on social exchange in delivering an intangible, there are greater opportunities both to manage and to mismanage it. At the moment the service is experienced, for example, both the provider and customer are generally nowhere near the immediate reach of management. One simple rule of thumb, therefore, is "the more people the customer must encounter during the delivery of the service, the less likely it is that he or she will be satisfied with the service."[17] The tangible economy taught people to live in hierarchies; the intangible economy is teaching them to keep the structure simple and flat, to stay close to the customer, and to pay more attention to quality than to authority.

Another attribute of intangibles is that the functional difference between producer and seller tends to be vague or nonexistent. People who deliver intangibles have to think holistically, integrating the production and sale into a simultaneous, single role that has not yet been formally recognized in management lexicon. Similarly, since customers are in a sense co-producers of intangibles, participating in the social and creative process, we must also develop holistic approaches to the producer-consumer interface.

The development of new models – coming from science and technology and applied to business and organization – will apply to entire companies, including both factories and offices, in all economic sectors, and both nationally and worldwide. New technologies enable each part of companies to be in simultaneous contact with every other part and with the whole; something that industrial hierarchies never allowed or desired. Because so much of our emphasis is on the office, it is useful at this point to address how the factory of the future reflects the same new modeling.

One source of new factory organization is in automation technology, particularly in what is called computer integrated manufacturing (CIM) and its miniature version, flexible manufacturing systems (FMS). As the names imply, computer systems enable all parts of the company to be linked simultaneously. Each part has its own computer-aided system, and all parts are linked and made into a whole by CIM. The process begins when the headquarters of a company decides what to make, based on market research, proprietary advantages in technology and manufacturing, financial resources, business vision, and a strategic plan. Then, the computer-integrated parts take over.

Computer-aided design (CAD) systems assist in the translation of ideas into the first steps toward material reality. This is where artificial intelligence will play an important role in linking human creativity and innovation to the new automated technologies.

Computer-aided engineering (CAE) systems take the next steps: they design the product, assure its quality and economic feasibility, plan the manufacturing process, design the molds and tools to be used, and program the production machinery to do the job intended.

From there, computer-aided manufacturing (CAM) molds, machines, welds, and otherwise fabricates raw materials into components ready for assembly. This is the workhorse of the CIM family.

Another CIM family member, computer-aided assembly (CAA) puts together these home-grown parts with other parts bought from outside suppliers. Then it automatically tests the products, and boxes them for (momentary) storage and shipment.

The automated warehouse, run robotically, moves raw materials, intermediary parts, work-in-progress, and final products to their in-house or market destination.

JIT (just in time manufacturing) is a cousin to the CIM family. With JIT, the bare minimum of these parts and materials need be kept in inventory. Because the producer will be able to deliver in real-time, nothing need be produced without the order in hand. This is another reason for the primacy of the marketing function, particularly sales and service, in the new economy.

Where are people in this brave new factory? Mainly, they are as absent here as they are absent from the automated farm. To the extent they are prevalent, it is for the asset of their brains, not the cost of their brawn. As with products, so too with people and organizations, their value will be measured in the ratio of the information they provide relative to the mass of their numbers. As knowledge workers, their jobs will be to run the entire process, coordinating the flow between the parts. These new factory workers, however, will not be the model for work in the new economy any more than farm workers in the then-new agribusinesses became the model for work in the industrial era.

The CIM family of systems automates and links all the factory functions to headquarters and to other parts of the business. In its complete form, many concepts of the industrial economy become obsolete. The learning curve, for example, would be eliminated from the least-cost producer model. The first item would be manufactured as cheaply and as well as the thousandth one. Because specifications can be changed instantaneously, there need be no lag-time in a company's response to changing market conditions. While in many lines, economies of scale in manufacturing will dissipate if not disappear, in R&D they will escalate. Companies will be able to customize their products and services on a mass basis, the subject of the next chapter. Current concepts of market-share will lose their meaning, and hence importance. Quick turnarounds, low inventories, and almost no labor costs would make it unnecessary to import goods from cheap labor countries. Factories would be located, once again, near the markets they serve. The automated factory is integral to the any time, any place reality of the future.

Those who have the new factories will significantly ease their entry barriers into new businesses. Those who don't will be seriously disadvantaged. The short-term costs of installation, however, are considerable. One FMS can cost as much as $5 million and take several years to install. Probably, both cheap-labor countries and small businesses in postindustrial nations will be severely pinched. Since three-quarters of U.S.-manufactured items come from shops with less than fifty employees, the coming of the automated factory will add to the demise of the manufacturing sector in the economy as we knew it. Despite the entrepreneurial flexibility often ascribed to smaller organizations, they will be hard pressed to adapt (let alone lead) this capital-intensive shift into the manufacturing future. Most of the early switches to the 2001 factory will come from only the very largest manufacturers.

According to the U.S. Commerce Department, in 1986 there were only 47 FMS's in the U.S., fifty in Japan, and eight-four in Europe. Despite the small numbers and large costs, however, Westinghouse is forecasting a $38 billion market in CIM by only 1990. Although this seems high, the CIM factory assuredly will be a more common reality by 2001.

At the same time that manufacturing is becoming more automated, networked, and holistically integrated, like all other parts of companies, it is also becoming less tangible. The physical reality of manufacturing, together with all other business functions, is less and less.

Industrial economies created corporations with large tangible assets. The factories and office buildings of major corporations have an imposing physical presence. The Chrysler building in New York and the Sears tower in Chicago are landmarks. Even the average Joe on the street knows a company by the name on the building. These physical monuments are the concrete (and steel and glass) representations of their corporations' importance. They have a tangibility about them that tells people of their presence. And even if one never travels to Detroit, we know that the General Motors headquarters building is there, together with many other G.M., Ford, and Chrysler factories and offices.

A phenomenon is occurring in the new economy, however, wherein the physical presence of companies is reducing to the point where it is difficult to say that they have a physical existence in any place. In the same way that a corporation is a legal being, yet you cannot literally touch it the way you can physically hold a human being, new forms of enterprise are evolving that are very real yet have no-matter. They have a palpable though non-physical presence. They are literally intangible companies.

In a cover story about this phenomenon, which it called "The Hollow Corporation," *Business Week* described one such company.

> *Lewis Galoob Toys Inc. is obviously a successful company. It sold $58 million worth of sword-wielding Golden Girls' action figures and other trendy toys last year — 10 times the 1981 total. Its stock, issued in 1984 at 10, has soared as high as 15 and now sells for 13.5. Yet by traditional standards of structure, strategy, and management practice, Galoob is hardly a company at all.*
>
> *A mere 115 employees run the entire operation. Independent inventors and entertainment companies dream up most of Galoob's products, while outside specialists do most of the design and engineering. Galoob farms out manufacturing and packaging to a dozen or so contractors in Hong Kong, and they, in turn, pass on the most labor-intensive work to factories in China. When the toys land in the U.S., they're distributed by commissioned manufacturer's representatives. Galoob doesn't even collect its accounts. It sells its receivables to Commercial Credit Corp., a factoring company that also sets Galoob's credit policy.*[18]

To farm is to produce (a crop on land); to farm out, however, is to send work from a central point to be done elsewhere. Farming out is not about farming, though we use a word from the agrarian economic period to describe the organization of activities that are taking place in an industrial context. Similarly, in an industrial context, factories are not about factors, though we use a word from the preindustrial economic period to describe the buildings where manufacturing takes place. The word "factory" comes from the medieval Latin *factoria*, meaning an establishment for factors. Factors are people or firms that accept accounts receivable as security for short-term loans. The factories of the new economy, such as Galoob Toys, may behave more like preindustrial establishments for factors than like industrial plants for manufacturing.

The value added into the economy by this new organizational form does not come out of giant industrial plants, even automated ones, nor out of giant offices of the future. Rather, it comes from a kind of organization that has little physical reality at all. Galoob, for example, does almost none of the traditional business functions itself. Research, design, development, engineering, manufacturing, marketing, distribution, sales, and finance can all be done by others.

All functions can be disaggregated and then subcontracted. This is exactly what is happening to large corporations. In their search for least-cost production, manufacturers increasingly turn to contract manufacturing and to the import of goods produced in foreign low-wage factories. This is especially true in electronics. As recently as mid-1984, for example, 60 percent of the employees in General Electric's consumer electronics business were in manufacturing. Within two years the number dropped to 10 percent.

A major advantage of this form is that the company can respond quickly to changes in the environment. By carrying low overhead expenses, and few fixed assets, it keeps its costs down. There are low entry barriers because it needs less capital. And it can tap low-cost labor and state-of-the-art technology easily, from the outside. On the negative side, however, it has less security of supply and less control over production. In time, such a company can lose its own expertise in design and manufacture. Then, it can be vulnerable to competition from its own suppliers. Without material investment it cannot cross-subsidize product lines, letting the proverbial cash cow pay for the rising star. Operating on no-matter, it is at once less rigid and less secure, more volatile and more flexible.

The factory of the future is clearly the automated, robotic environment with relatively few employees. More relevant, however, is that this factory of the future will be only as important to the economy of tomorrow as the farm is important to the economy of today. That is, it will have an essential place, but with very few people adding a smaller and smaller proportion of the value to the total economy.

Perhaps a more useful way to think of the shift is to see it as paralleling the shifting importance in information technology from hardware to software. The automated factory of the future is a factory for hardware. The increasingly important factory of the future, however, is represented by the software factory. This is only partly because it is in the central field of the new economy, information processing. More important, it is because it represents the research, design, development, engineering, manufacturing, distribution, marketing, sales, service, and administration of intangibles, the increasing economic value of no-matter.

It is not coincidental that software shops tend to be small, brain-intensive places, built with low fixed investment, and able to change quickly. The factory of intangibles, rather than the automated hardware factory, is the more likely candidate for the organizational model of the future. And the place is more likely an office than a factory.

Debates about the merits and demerits of a no-matter economy would not be complete without consideration of its productivity. Saying that something is immaterial literally means that it has no tangible body or form; more usually we mean that it is of no consequence or importance. Services are literally immaterial, and though now no one would argue that they are unimportant, services are still demeaned for lacking the productive contribution to the economy made by their industrial sisters.

A major problem is how to define productivity. Grossly, productivity is some measure of output divided by some measure of input. For industry, these are generally sales divided by either labor costs or capital invested. Such measures fail to account for quality or value, crucial factors in service delivery. Sales per employee is a better measurement, though not ideal; and, using this index, Fortune finds the largest service firms more productive than the largest industrial corporations.[19] A more relevant measure, though still fairly unrefined, is that of value added per employee. Here, the value of goods and services purchased by the firm are subtracted from sales, and the remainder is divided by the number of people employed. Since service companies purchase fewer intermediate items than manufacturers, according to James Heskett, "it is quite likely that the value added per employee in services is actually *greater* than that for manufacturers."[19]

The debate over productivity has obsessed policy makers and managers for years. It is argued about both in theory and in practice, both in macro and micro terms.

It is seen as the major reason for our past economic decline and also as the promise of our future economic growth. In even the most optimistic scenarios, however, the low productivity of our services is the most worrisome element.

Since the Bureau of Labor Statistics forecasts that nine out of ten new jobs during the next decade will be in services, and since the average hourly pay in services is 11 percent lower than in manufacturing, there is widespread concern about the continued slide in real standard of living. Chase Econometrics, for example, predicts that per capita income will grow only 1.5 percent per year, less than inflation and half of GNP growth, mainly because of the income-generating drag of services.

Even using the artificial, inaccurate, and outdated dichotomy between industry and services, however, the gap in productivity between the two is clearly diminishing. Part of this is because of our increasing awareness that we don't have an adequate framework for distinguishing between service and nonservice economic activities, or for measuring adequately what constitutes productivity. In addition, just how productive services are depends on whether one is taking a snapshot measurement at one point in time, or a moving picture of service productivity through time. Despite real concerns about the false paradise of a service economy, there are good reasons to believe that the return on investment in service technology will clearly be felt by the turn of the century.

Using Bureau of Economic Analysis data, Stephen S. Roach at Morgan Stanley & Co. calculates that service providers owned $126.1 billion in new technology in 1975. Adjusting for inflation, that figure jumped 95 percent to $246 billion in 1982, and 46 percent to $358.9 billion in 1985. Since white-collar employment rose 42.4 percent between 1975 and 1985, the result is a doubling in new technological investment per white-collar worker during that time. That amounts to about $7,558 in 1985, or 10 percent per year.

Charles Jonscher of M.I.T. estimates that "each $1,000 invested per service worker in the new technology is twice as productive as the same $1,000 invested per industrial worker in machine tools or conveyor belts."[20] Because of the educational impact of investment in new technologies for service, Jonscher expects that by 1990 the productivity surge in services will be four times as great as for the same amount of industrial investment. Although we may debate perennially how to measure productivity, there is little doubt that by 2001, when we are six decades into the service economy, services will be as productive as industry was in 1921, six decades into its economic epoch.

Another gloomy side of the debate on the productivity of no-matter is the expected early arrival of "smokestack services," due to the combined effects of deindustriali-zation and globalization. Critics argue that tangible out-put from some postindustrial economies shrinks as that output moves to low-wage countries. The atrophy of manufacture is not driven by the growth of the service sector. The manufacturing sector has been in decline in the U.S. since the 1940s. Now, there is reason for con-cern that the same thing will happen to intangibles.

For a long time the U.S. advantage in its export of intangible services helped to balance the mounting trade deficits in manufactured tangibles. Now, competition on worldwide trade in intangibles is catching up.[21] The U.S. still holds a commanding, though narrowing, lead in advertising, finance, software, and television. Other sectors, however, such as aviation, leisure, shipping, travel, and tourism are beginning to resemble our declining smokestack industries. The stakes are big; world trade in intangible goods is estimated at over three-quarters of a trillion dollars.

Our competitive edge is fragile, according to *Business Week*.

> *If profits earned from foreign investments and repatriated to the U.S. are excluded, the American service champion becomes an also-ran. By that measure, Europe exports three times more services than the U.S. And if "smokestack" services such as tourism and shipping are excluded, the U.S. ranks fourth in exports behind Britain, Germany, and France.*[22]

In the ten years 1973–1983, the U.S. share of global trade in business intangibles dropped from 15 percent to 8 percent.

For example, the U.S. has lost its dominance in shipping just as it has lost it in shipbuilding. Countries like Taiwan have combined low labor costs with computer-navigated, fuel-efficient ships that require only small crews. The fleet capacity of Taiwan's Evergreen Marine Corp. is around 100,000 twenty-foot container units, making it the world's largest container line. Sea-Land Industries USA, which pioneered container shipping, is struggling way behind.

The U.S. share of Mideast construction contracts dropped from first to twelfth place during the first half of the 1980s. Japanese and Korean firms have been the big winners, forcing giants like Bechtel to enter into partnerships with new service giants such as Korea's Hyundai Engineering and Construction Co. Saudi Arabia canceled an exclusive $1.4 billion hospital management contract with American Medical International, opening the way for European and Asian competition to challenge the market leadership of U.S. health care companies.

Brazil and India, as well as Britain and France, are moving strongly into software exports. Japan, too, is brooking the language difficulties, and bringing their quality control to bear on software and on the ultimate intangible, artificial intelligence. Korea's Daewoo, Gold Star, and Hyundai are also moving to become major forces in software export.

Started only in 1970, Saatchi & Saatchi became Britain's largest advertising agency in 1979, Europe's biggest in 1981, and with the 1986 acquisition of the larger U.S. ad firm, Ted Bates, it became No. 1 worldwide, with combined billings of over $7.5 billion. They now employ 12,000 people in twenty-eight countries. Its U.S. businesses provided 58 percent of 1985 profits, and that was even before the Bates deal, and a few others in the U.S. Nor are they stopping at the globalization of advertising. They are also taking aim at consulting intangibles on a worldwide basis.

In 1985 Saatchi & Saatchi acquired The Hay Group, a compensation consulting conglomerate with 100 offices worldwide, and the Yankelovich, Skelly & White market research firm. This was enough for their consulting businesses to produce 29 percent of 1985 profits. They are shooting to make that figure 50 percent by adding major acquisitions in strategic planning, pension and actuarial, and financial systems consulting. By far the most aggressive ad firm, they are using the intangible resources of creativity, ambition, and financial genius to redefine how businesses that deal with intangibles, such as advertising and consulting, will have to compete worldwide in the future.

European companies such as Reuters Ltd. already dominate the financial information market. Many are buying up U.S. business information services, such as Wharton Econometrics, which was bought by a French company. According to Link Resources Corp., non-U.S. companies own one-third of the almost $2 billion on-line database market.

If the trend continues, by 2001 it is possible that the center for financial service exports will have moved to Tokyo. Japanese banks are now firm lenders to U.S. companies, providing almost 10 percent of commercial and industrial loans. Millions of dollars in government-backed mortgages have been sold in the Tokyo capital market, meaning that Japanese savings are helping to buy more and more homes for Americans. Japanese insurance companies have also expanded overseas, using domestic profits to subsidize their entry into the intangibles trade abroad. Tokyo Marine & Fire Insurance Co., for example, is the world's largest marine insurer and Japan's largest non-life insurance corporation. 17 percent of its premium income comes from operations in twenty-two countries.

In sum, the new service exporters deal with intangibles on a worldwide scale. They depend on creativity, information, communication, and distribution in an any time, any place world, more than their industrial older brothers who produce material goods in fixed time, space, and, often, costs. As global competition increases in intangibles, as it did in tangibles, the job export shift will move into services, following a similar pattern as it did in industry. The first jobs to go will be the ones that add least value to the new economy, and that can be done any time, any place. Clerical and data entry workers are likely candidates by the turn of the century. As the no-matter economy matures, the spectre of smoke-stack services appears and the U.S. lead in intangibles is assaulted by global competitors.

The matter/no-matter split is deeply rooted in Western thought, which regards matter as the opposite of mind and spirit. By contrast, with the new economy's emphasis on the intangible, we must develop approaches that regard the two once again as equivalences. This has happened in science and technology. It is just beginning to be appreciated in business, with the growing awareness of the service economy. In terms of management and organization, however, it is hardly yet thought about. In science, you just have to get small enough – sub-subatomic – to locate mind in space. The journey is worthwhile, however, because mind is a limitless resource, and probably the most critical one in the economy of 2001.

For the transformation of management fully to take place, it is going to require major shifts in the way we think. Transformations in scientific thinking have their analog in management thought. In mathematics, for example, simple equations require us only to think sequentially. To solve simultaneous equations, however, requires us to think in a holistic context, of all parts and the whole simultaneously. In business, services operate much like simultaneous equations, in that all properties must function simultaneously to work properly.

Or take another example: You need two numbers to locate a place in two-dimensional space, three numbers to locate a place in three-dimensional space, and n numbers to locate a place in n-dimensional space. This is so, whether we are talking about the universe, the marketplace, or a box in an organization's structure. Matter can be fixed in time and in space. No-matter cannot be fixed this way. It exists, yet in no particular space.

How many numbers does it take, therefore, to locate the place in space of a nonmaterial "thing," that is, of no-matter occupying no space? In a decade or so, when we get well enough beyond the hardware phase of the office-of-the-future revolution, won't we have to conceive of the organizations running these offices with a comparable n-dimensionality? These kinds of questions are difficult, if not impossible, to answer with traditional models. They require a new model in which any number will do because no-matter "things" exist in every place at the same time, what Eastern philosophers call a oneness.

This is a difficult concept for Westerners to grasp. Most have rejected it (too soft and intangible?). Still, it is making its way into our models of management through what we consider the more "solid" route of science and technology. It is a common statement in the computer companies that their business is at the point today where automobiles were in the 1920s. The infrastructure is barely in place, and still evolving. There are islands of intense penetration and heavy usage, but they do not yet affect everyone directly, and they have not yet been connected into a complete network. Their impact will go far beyond the machines, and will affect the very fabric of society as well as the economy. They will change the way we think and act, but not overnight.

Managers want answers overnight. They are not very interested or concerned with what will be one and two decades away. But as with all great leaders and leading corporations, the best participate in creating the future, making it happen sooner. In every business, a key to the future happening today is for its management to perceive the power of a different way of looking at their world and business *before their competitors do*. Managers who can deliver their products and services any time, any place, relative to their competition, have a decided advantage. Similarly, the competitors with an important edge will be the ones who can identify and manage the value-added by the intangible core of their offering, and who can increase the ratio of that intangible/tangible value better than their rivals.

The process begins with a transformation of scientific thought, with new perceptions about how the world works. These are translated into new technologies, practical applications that take form in products and services. In industrial economy the form is generally tangible. Today, we are awakening to the fact that the form is basically intangible. It has taken science and technology decades to explore the abundant significance of this truth. Business management has only been at it for about a decade. It has barely formulated the correct questions. Managers who want to know the future of their business and organization would do well to look to science and technology for predictive clues. The transformations that occur in one realm also take place in the others.

Mass
Customizing

We are all unique.*

*This statement is false.

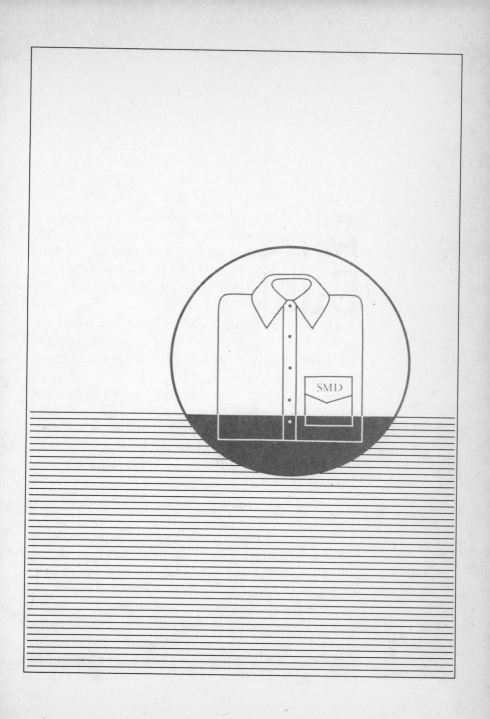

As the new economy matures, many new concepts, theories, models, and frameworks will develop that are appropriate to actual conditions, not holdovers from the industrial economy. One seems particularly ready to make its debut now: mass customization.

Let's begin with mass customizing shirt manufacture. Any shirts produced at one time and with the same specifications are parts of a single whole production run. Producing one custom-tailored shirt means the whole has only one part; the production run in a factory may mean the whole equals 5,000 identical parts. What if technology made it possible for every one of the 5,000 shirts to be customized while on the factory assembly line — that is to say, produced just as quickly as the 5,000 identical shirts, yet at no greater expense? Each shirt then is both a whole and a part of a whole at the same time.

The world of mass customizing is a world of paradox with very practical implications. Whether we are dealing with a product, a service, a market, or an organization, each is understood to be both part (customized) and whole (mass) simultaneously. New technologies are now coming on-stream which deal with infinitesimal parts of the wholes that interest us. They are able to get specific about parts that earlier technologies had to leave undifferentiated. In addition, they operate at such fast speeds that we may consider their treatment of parts simultaneous. Speed and specificity are the hallmarks of these new technologies, and the foundation for the mass customizing of products and services that follow. Speed and specificity enable us to see how the whole is actually present in each one of the parts.

For mass customizing of products, markets, and organizations to be possible, the technology must make it economically feasible in every case. One of the major propositions of this book is that the models we use for managing and organizing corporations have their antecedents in the product-market relationships that we call businesses, and that these, in turn, are premised on the new technologies. It is logical, therefore, that new technologies will display abilities that later find their way into business and organization models.

In this section we will look at what five particularly interesting technologies have to offer us about new models for the new economy: holography, parallel processing, customized chips, biogenetic engineering, and customized catalysts.

HOLOGRAPHY. The principle of holography was spelled out in 1947 by Denis Gabor, who later received the Nobel Prize for his effort. Like many scientific discoveries, however, it had to wait a number of decades until an appropriate context developed in which to utilize its potential. The context was the invention of the laser, which made possible construction of the first hologram.

A hologram is a kind of 3-dimensional "picture" produced by lenseless photography. As described by biologist Lyall Watson, "When two laser beams touch, they produce an interference pattern of light and dark ripples that can be recorded on a photographic plate. And if one of the beams, instead of coming directly from the laser, is reflected first off an object such as a human face, the resulting pattern will be very complex indeed, but it can still be recorded. The record will be a hologram of the face."[1] The holographic picture is a 3-dimensional image, not on the photographic plate but projected into space, in much the same way stereo sound seems to come not from either speaker but from a space in between.

For our purposes, the hologram has a very unique property. *If the image is broken, any part of it will reconstruct the whole!* This is important to mass customizing because, as Marilyn Ferguson says, it means that "the whole code exists at every point in the medium." Each (different) part of the image embodies the whole, much as each shirt in a customized production run is both a whole and a part of the whole simultaneously. We will return to the implications of this powerful and puzzling reality at the end of this chapter. Here, we must move from the scientific discovery of how part (whole?) of the universe operates, to the impact of its technological applications in business and organization. There are three main areas where holographic application has been the strongest: interferometry (testing and analysis), display holography (3-dimensional imaging), and holographic optics (light manipulation).

Holographic interferometry is a technique that permits measurement to an accuracy of ten-millionths of an inch, smaller than the wavelength of light. Changes can be detected in a concrete slab, for example, caused by the weight of a paper clip. This sensitivity allows engineers to "see" inside objects at the molecular level and custom-tailor infinitesimal parts. According to H. John Caulfield, of the University of Alabama's Center for Applied Optics, "by holographically recording an object, changing it slightly through application of heat or pressure, and re-recording a hologram of this second state of the object, we can experience two objects occupying roughly the same space but occurring at totally different instances in time. Fundamental laws of the physics of space and time are seemingly suspended. The effect is astonishing."[2]

Many fields of knowledge are limited: they know *what* cause will produce *what* effect, yet they are unable to tell us the actual process by which the effect takes place. The information provided by holograms can tell us *how* various phenomena occur. There is far greater leverage in knowing *how* than in knowing *what*. Each new understanding leads to multiple applications. To take but one, visualizing the shockwaves of an object moving through space, Caulfield says, holograms can tell us "how fuels burn, how air currents flow around automobile and aircraft parts, how effectively aerosol spray nozzles and ignition systems are designed, how mosquitoes fly, and how the core of a nuclear reactor dissipates heat."

Display holography is used increasingly in advertising and entertainment, but its greater value is in science, medicine, in conjunction with computers, and for identification and security. Scientists can image weather patterns, ocean currents, topographical maps, petroleum deposits, and cell growth. Doctors transform consecutive X-ray images, made with a CAT scan, into a 3-dimensional holographic skeleton. Creating 3-dimensional images from 2-dimensional data requires using holography in conjunction with computers. Among other things, engineers can use this technology to previsualize and design parts for cars, planes, and anything else. The now popular car advertisements that show CAD graphics of the car are still only 3-D simulations projected onto the 2-D space of the television screen or the magazine page.

Imagine (image) how different our world will be when all of these are presented in actual, not simulated, 3-D. Television will be 3-dimensional holographic stereograms; and the businesses that build the sets and produce and transmit the programs – as well as the organizations that run these businesses – will ultimately be just as transformed.

Display holography is also an inexpensive and foolproof way to store, conceal, identify, and secure information. Companies issue millions of credit cards, each one pretty much like the other, also very much like the millions of fraudulent copies that circulate as well. In 1986, over 90 million holders of MasterCards had a hologram embossed on their cards and became intimately familiar with the eerie 3-D picture. At a wholesale cost of about 2.5 cents each, the hologrammed card will be impossible to copy or alter without destroying it. Passports and currency may also be protected by technology. Customized holographs may replace the card customizing which is currently secured by one's signature and three other forms of identification.

Holographic optical elements (HOEs), are lighter and less expensive than glass lenses, and they can be customized to fit any shape or light wavelength. HOEs, for example, are already operating in many supermarkets. As the packaged groceries are passed over the checkout counter window, holographic lenses on a spinning disk direct laser beams to "read" Universal Product Codes. Transparent instrument panels, connected to computers, can display information as if it were floating in the window of a vehicle, so that the driver or pilot need not look down at the dashboard. Holographic lines in space zero in on military targets or fogged-in runways, and similar applications can be used in other forms of alignment and surveying.

Holographic applications of the basic scientific discovery have scarcely begun. Until the technology brings myriad products on-stream, it is not reasonable to assess what will be the impact on businesses per se, let alone on the organizations that will run those businesses. Since the impact will most certainly be felt, however, it is not unreasonable to pose two questions: What does a business look like when viewed holographically? What is a holographic organization?

PARALLEL PROCESSING. The scientific and engineering principles upon which the computer was invented were outlined by mathematician John von Neumann in 1946. In his approach, a computer's central processor gets its instructions and its data one piece at a time from a main memory, pausing after each piece to send the results of the process back to the memory core. From the human point of view, this all happens very fast, but for electronic systems that measure time in billionths of a second, the processor has to suffer many delays waiting for the information to come and go.

The elimination of even these time delays, the "electronic downtime," has been an engineering goal for decades. During that time engineers accepted the step-by-step sequence in the von Neumann principle and tried to minimize the problem by miniaturization. Shrinking the size of the transistors on the computers' integrated circuits and packing them closer together helped, by reducing the distance the information had to travel during the sequence. Like cars in a traffic jam, however, the more packed in they are, the more heat they generate. Ultimately, the wires melt.

The solution to the problem requires a contextual shift, away from a step-by-step sequence and toward a simultaneous approach, where many processors handle different parts of a job at the same time. This is the world of multiple or parallel processing.

Multiprocess computers work at extraordinary speeds. Processing work can be measured in terms of the number of (floating point) operations per second, or "flops." "Megaflops" are millions of flops, and "gigaflops" are billions of flops. In 1970, the ILLIAC IV handled fifty megaflops, and a decade later Control Data built the Cyber 205 to handle 400 megaflops. Cray Research Inc. is breaking speed barriers even faster with the Cray-2 peaking at 1.2 billion flops in 1985 and the Cray-3 at 10 billion around 1990. IBM is also building a ten gigaflop supercomputer. For those who fear the Japanese onslaught in superfast computers, their Ministry of International Trade and Industry (MITI) has targeted the same 10 billion for 1990. The Advanced Research Projects Agency of the U. S. Defense Department, however, is driving at a trillion operations per second by around 1992.[3]

Quantitative advances of this magnitude create qualitative shifts. Initially, these shifts will be used to tackle more complicated problems in basic research, including such technological breakthroughs as the development of fusion energy and a bioengineered cure for cancer. Faster computers will make possible structural analysis of asymmetrical molecules and of the motion of molecules in general. More practical will be the development of better car engines and safer airplane designs. By 2001, superfast computers will be doing mundane operations, such as customizing products for people on a mass basis.

There are only a little over 100 such supercomputers today, although estimates range between 200 and 500 shipped per year by 1990. Furthermore, parallel processing will move down to mainframes and smaller machines as well. Michael L. Dertouzos, Director of the Laboratory for Computer Science at M.I.T., believes that within a decade personal computers will have over 100 processors in them. When this happens, everyone will begin mass customizing – and demanding it from others. Then the bottleneck will once again be software, since all software is currently set up for serial, not parallel, programming.

CUSTOMIZED CHIPS.   As with many technologies, yesterday's state-of-the-art black box is today's commodity. Developments come on-stream so fast that the shortened product life cycles often make it difficult to recoup development costs. One way to do that is to pass those costs on to the buyers, but the latter will only be interested in absorbing these costs if the product is tied to their unique needs. Addressing the specific needs of any one chip buyer sounds like an expensive proposition, yet that is exactly where the industry is headed.

In an industry that is cyclically plagued with overcapacity in most standard products, chipmakers will increasingly move toward semicustomized and customized integrated circuits. While the standard off-the-shelf chips will always have some place, they will be the increasingly unattractive bottom end commodities, supplied only by super-efficient least-cost producers. When so many transistors, or circuit switches, are crammed onto the microspace of a computer-on-a-chip, they will have to be tailored for specific uses to fulfill their potential. To amortize the engineering costs of such complex chips will require bringing the user directly into the design stages, creating customized chips that are produced en masse.

When software development groups start off on their own, they are likely to develop software that the user cannot use. The same is true for the hardware design of the custom chip. This need to be physically close to the end user, therefore, will draw the large-scale customers into the manufacturer's space for the customizing process. Although the semicustomized and customized segments represented only about 12 percent of the market in 1985, it is expected to at least double by 1990, and need grow only 10 percent a year to virtually dominate the market by 2001. To keep up with Japan, U.S. manufacturers will have to move into the mass customizing of computers-on-chips.

Intel, Motorola, NCR, RCA, and several specialty shops produce such integrated circuits. They are made on silicon wafers 6 inches in diameter, and yield twice as many chips as the earlier 4-inch industry standard, with little added processing cost. Even more important are the engineering costs. An industrial model might figure out ways to pass the cost on to the user; however, the model for the new economy increasingly enlists the *consumers'* direct participation in the value-added steps of R&D, engineering, manufacturing, and/or distribution. In the case of customized chips, the participation is in the engineering. New computer-aided design tools make it possible for customers to write their own specifications directly onto the chips that others produce for them, binding the customer closer to the manufacturer in the process.

Not only will no-matter specifications routinely be written on chips, but, in the future, the atoms of the chip's material will also be tailor-made. AT&T Bell Labs has been able to transplant a clump of just 10 atoms, reshuffling them to produce precisely the semiconductor properties wanted.

BIOTECHNOLOGY AND GENETIC ENGINEERING. Another field where technological breakthroughs increasingly permit mass customizing is biology. Mendel formulated in the nineteenth century the theoretical principles of genetics. From these we know that genes determine heredity, though the structure of the genetic mechanism was not discovered until almost a century later. The current biological revolution probably began with the discovery of the double helix by Crick and Watson in 1953. Since then, biologists have understood *how* biological characteristics are transmitted, and they have proceeded to engineer life forms by acting directly on the genes. In terms of our basic progression, biotechnology and genetic engineering are about to move from technology to business, from R&D to (customized) mass production. A major part of this development are the highly customized and mass produced miracle drugs.

In agrarian economies, the local healer, or a household member, would custom-blend a batch of natural herbs and medicinals to heal a sick person. Industrial economies, by contrast, learned to produce healing drugs in enormous volumes. Remembering that the U.S. industrial economy spanned 1865–1945, however, the early drugs that were mass produced were very multipurpose. Patent medicines, as the joke was told, would cure most everything that ailed you. As the modern pharmaceutical industry grew, it produced more specific medicines for specific diseases, and in very large quantities.

The earlier generations of remedies were woefully hit-or-miss, as were the methods for discovering them. Finding a new drug pretty much meant trying tens of thousands of chemicals until you hit on the right combination. Even then, the new drug generally attacked the disease, not the fundamental cause that lay more deeply buried in molecular biology.

Today, however, scientists are able to get at the specific substances – the enzymes, hormones, and factors regulating cell functions and pathways. One or more of these specific substances is deficient in a diseased person, and molecular biologists assume that, if there is activity (or isn't activity where there should be), there is a gene controlling that activity which can be identified. Once identified, specialized enzymes snip out the relevant pieces of DNA. Substances that occur in infinitesimal quantities in the human body can then be produced in large amounts. These scientists virtually walk up and down the gene looking for sites to engineer. The path begins in the micro-universe, where breakthroughs in science lead to technological applications, and ultimately to the mass customization of drugs – new products for the industrial sector of the new economy.

A prime example of the mass customized drugs are the magic bullets known as "monoclonal antibodies." Antibodies are the first line of defense against infection. In the 1970s British researchers discovered a way of fusing an antibody-secreting cell and a cancer cell. The outcome produced large quantities of the ultrapure, or monoclonal, antibody that seeks out and locks onto a specific target, killing only the cancer cells without harming healthy tissue.

Drug makers are now testing large numbers of such products. Johnson & Johnson has an antibody that blocks white blood cells that destroy transplanted organs. Other companies are taking aim at a variety of cancers, multiple sclerosis, and "hospital acquired infections" which cause 80,000 deaths annually. Multiple sclerosis is caused when a specific kind of white blood cell, that normally attacks bacteria, goes haywire and strip the protective sheaths of nerve cells. The magic bullets destroy these specific white blood cells.

The first vaccine, against smallpox, was developed in 1778 by Edward Jenner. Although there have been only ten widely used vaccines since, the mass customizing of drugs has produced about twenty others that are currently being developed. Among those that may win FDA approval by 2001 are vaccines for AIDS, chicken pox, cholera, croup, viral diarrhea, dysentery, gonorrhea, infectious and serum hepatitis, genital herpes, malaria, meningitis, rabies, strep throat, typhoid fever, and tooth decay.

Traditional vaccines inoculated the person with the killed or weakened virus or bacteria, causing the recipient to form antibodies. The problem was that it also occasionally caused the disease it was intended to prevent, and sometimes had side effects that were worse than the disease itself. The new vaccines, however, are so targeted that they significantly lessen these dangers.

There is even talk of a cancer vaccine. Today, the best cancer treatment is to catch it early and remove it surgically. But attempts are now being made to synthesize, and produce in quantity, little-understood substances of the immune system. Research at Genentech is attempting to bond gamma interferon, which seems to suppress tumorous cell growth, with TNF (tumor necrosis factor), which seems to destroy tumors already there. Ultimately, cancer prevention may come from the other route to mass customization. That is, isolating the specific gene that may trigger the uncontrolled growth of tumors and develop monoclonal antibodies in large supply.

The truly significant drugs that are expected by 2001 will come from "protein engineering," that is, modifying specific chemicals of the body, then mass producing them as synthetic compounds that selectively block disease processes. Because they are synthetic, they cannot be broken down by the body; therefore, they can be taken orally because the digestive system will not destroy them.

The technological key to customized molecules lies in computer modeling that also uses holographic techniques. Sophisticated computer graphics, like those developed at the National Institutes of Health, instantaneously show researchers how a drug reacts with a body protein at the molecular level. With these kinds of tools, over the next decade, molecules will be designed the ways cars and jet engines are designed today. The result will be customized molecules, mass produced to fight and prevent disease.

None of this will come cheaply. Labs for the new drugs now cost as much as $100 million to build and $75 million to run each year. This causes major pharmaceutical houses to double their R&D expenditures, exceeding $4.5 billion in 1985. With these kinds of costs, many biotech firms will not survive to see the millenium. Those that do, however, will be mass customizing miracle drugs.

CUSTOMIZED CATALYSTS. $750 billion worth of products, almost 25 percent of our gross national product, are manufactured each year with the aid of catalysts. Even in its modern phase, catalysis has been a rather black art. Developments in recent decades are still based on a trial-and-error approach. Chemists know *what* happens in a shift from state A to state B, but they don't know *how*, the specific mechanisms by which it happens.

If scientists can understand the mechanisms by which catalysis takes place, they can custom-design catalysts to produce specific chemical products. This is precisely what is beginning to happen. Enzymes, natural catalysts, tend to be slow, but they are very selective in the conversion of raw materials to chemicals. The central aim of the new catalytic chemistry is to *design* such selective catalysts rather than *discover* them.

Custom-designed catalysts will be able to mass-produce a myriad of custom-designed products. Catalysts-by-design can be used to make chemicals and all man-made fibers, drugs and vitamins, fertilizers, herbicides and pesticides, plastics and adhesives, and foods and fuels.

The petroleum business is a good example of the

impact this shift will have. Oil and gas are both composed of hydrocarbons. The hydrocarbon molecules are tightly packed together in oil, compared to gas, making it much easier to pipe or ship to refineries for processing. To ship the gas it has to be reduced in temperature until it becomes liquified. Liquified natural gas (LNG) is both expensive and dangerous. However, there is a lot more gas lying around fallow than there is oil.

Finding a safe and economic conversion process for transforming the gas is therefore a major goal, and the key seems to be in conversion chemistry. The direct oxidation of natural gas, methane, to methanol is possible within the next five years. If natural gas in remote and hostile locations could be converted to methanol at the well site, the liquid could be shipped far more conveniently than gas, which demands a costly high-pressure pipeline or LNG tankers.

During the past five years, for example, scientists have modified catalysts called zeolites. According to a catalytic chemist who worked for Exxon, "these crystalline forms of silicon aluminum oxide have well-defined molecular geometries, or tunnels that can be adjusted to guide a particular reaction. By design, the zeolite's tunnels will admit only molecules of a certain shape and size, rejecting all others. The reaction takes place, and the product molecule is determined by its ability to fit and navigate these tunnels. One such zeolite catalyst will be used to selectively produce high-octane gasoline from natural gas-derived methanol in only one step. And a modified version of the same catalyst has been developed to make a new chemical building block for producing an improved plastic."

Shell, with its reputation for outstanding technology, is attempting to convert natural gas into gasoline in Malaysia, by applying a variant of the so-far uneconomical Fischer-Tropes process. If you can make gasoline out of gas feedstock in Malaysia, you can make other things out of it also – things that add value through conversion close to the source, and that are more easily transportable than gas. Custom-designed catalysts are key to the success of this approach, and they require a design-, rather than a discover-, mindset.

The Fischer-Tropes process, which produces a synthetic gas of hydrogen and carbon monoxide, works in three steps. The first makes the building blocks, while the second and third put the blocks together to make new products. Fischer-Tropes is used for the first step, the front end, where 80 percent of the costs lie. Yet, everyone seems to be working the back end, inventing new products. By this reasoning, even a 10-percent improvement in the front end means an 8-percent reduction in costs. The new chemistry allows you to work with the front side of the chemical reaction where there is more meaningful change. The next decade may produce a race at the front end.

At the upstream end of the business in the future, increasing amounts of oil will be recovered from second, third, and fourth crops and from technologically assisted first crops. This means that the oil production business is going to be higher-tech than in the past, and will need to know *how* and *why* the recovery technology works, not just the *what* of it. Downstream, the new catalytic chemistry processes promise (1) cheaper production techniques in the bulk commodity products which will give greater leverage, and (2) a host of new product possibilities in the specialty businesses.

The shift from discovery to design will also create major organizational issues for petroleum companies. The historical mindset in the industry is to look for, find, and produce oil and gas. This is an orientation that discovers, not designs, oil and gas. With the long-time cycles in the oil industry, companies could afford to rely on a scientifically guided, trial-and-error orientation, and access experience to shorten the search. With the new catalytic design technology available, however, lead times will shorten radically.

The background of petroleum people tends to be mechanical and civil engineering with oil field experience, not process chemistry. The latter group needs to be protected, nurtured, and developed; otherwise they will get swamped by the dominant petroleum culture. A parallel example is how a data processing mindset stunted the word processing side of IBM until the latter group were structurally separated. By then, IBM lost years in entering the personal computer market.

Being able to accept the paradoxes of the universe is a major asset in making scientific discoveries, and also in applying them to technological inventions and innovations. We have examined the paradox of the simultaneity of opposites, particularly of the whole and its parts. We have seen how this pertains to some specific technologies, and to how it facilitates mass customizing. The same is true for advanced ceramics, CAD/CAM, expert systems, fiber optics, lasers, polymer composites, robotics, and a host of other new technologies.

The technologies make mass customizing a practical possibility, yet they are not sufficient by themselves. What is also needed is the perception of the mass customizing principle. In some businesses, this perception is sufficient without major new technologies. The new technologies do, however, facilitate the widespread application of mass customizing to many more products and services than might otherwise be the case. In this section we will look at mass customizing products and services, rather than at the specific enabling technologies.

I happen to have big feet, and with the dollar very strong against the pound in 1984, I ordered a pair of custom-built shoes in London. I was told at the time that they would be ready in six months. I might as well have been in an agrarian economy. I usually buy a mass-produced pair from the limited selection made in my size. Then there is no wait at all. From the customer's point of view, I can purchase a customized product if I am willing to wait six months, or a standardized one for about the same price immediately. One shoe industry expert, however, says that laser-cutting and robotics technologies will make it possible to customize shoe lasts at reasonable costs, and produce a made-to-order pair of shoes in your neighborhood shoe store in a matter of days.

Similarly, because I am 6'7" (2.02 meters), it is hard to get shirts that fit. American shirt makers give you long enough sleeves, but they fail to lengthen the body of the shirt proportionately. My solution is mass customized. Every three months a Hong Kong tailor visits the local motel in my neighborhood, and for the same price as the standardized store-bought variety, I buy made-to-measure shirts. They even have my initials on them instead of Yves St. Laurent's, at no extra cost. The waiting time is six weeks.

If you go to Hong Kong, you can have the shirts made in twenty-four hours. In Japan, 60 percent of men's suits are sold door-to-door. The salesmen, employed by department stores, carry ten sizes for fitting, and they customize by color, style, and material. To varying degrees, these are all examples of any time, any place mass customizing. But there is still a delivery delay.

Clothing manufacturers can approach mass customizing by the way they mix fabric, style, and color. One shirt maker, for example, may produce shirts in one fabric, with ten styles, and ten colors, while a second produces shirts in one fabric, one style, and one hundred colors. Both manufacturers generate one hundred stock-keeping units (SKUs); but, while the customers' perceived choice from the first company is ten, it is one hundred from the second company. When Ralph Lauren began, he offered basically one fabric (cotton interlock knit) in one style (short sleeve, one type of collar, logo) in a burst of shades. Later, scale enabled him to add a second fabric (pebble, or birdseye knit). Benetton follows the same logic.

Mass customizing does not always work, however; sometimes it runs into cultural barriers. In Italy, for example, a clothes manufacturer told his retailers that he could supply customized suits on a mass basis, by adding 10 percent to the price and a fifteen-day wait. In six months he sold only 250 suits that way, versus 200,000 the standardized way. His explanation for the failure is that Italians buy fashion and don't want to wait before they can wear their purchase. That is why all pants are sold finished in Italy. In France, however, pants are sold unfinished, because 80 percent are sold in small stores that can't carry large inventories. The final customizing touches are put on by the boutique owners who double as tailors.

There is an increasing trend for manufacturers and distributors to mass customize their merchandise. As in clothing, some of the customizing has been going on for decades; what has been changed is the elimination of the wait. CAD/CAM, for example, makes possible instantaneous changes in the specifications, customized adjustments in garment cutting without any machine downtime. The general message is, the more a company can deliver customized goods on a mass basis, relative to their competition, the greater is their competitive advantage.

The Cabbage Patch Doll craze of a few years ago was

another example of the mass-customizing application. I don't know about you, but I didn't think that these dolls were all that terrific. What made them so popular? The answer is that, just like our own children, each doll was unique; each was different from the rest, even though they had all been mass produced together. The customizing lay in the CAD/CAM signal to make ever-so-slight variations in each doll, and in the adoption paper that came with each one. One mass-customizing touch of technology, and another of marketing, added up to a multimillion-dollar-business, selling fairly regular dolls at premium prices.

The Japanese housing industry has apparently learned the mass-customizing message much better than their U.S. counterparts. Mass-produced, factory-built houses are not highly esteemed in the U.S. The general sense of the public is that what you save on costs you lose in quality and in customization. If you are to buy a mass-produced house, your choices are limited to gross differences, such as the two- or three-bedroom model. The Japanese, however, have applied mass customizing to the housing business. Their effort began in the 1960s and has become quite sophisticated since then.

The process begins by sitting down with a sales representative for a couple of hours and pretty much designing your own home on a computer screen. You choose from 20 thousand different standardized parts, like life-sized Lego blocks, and you can put those parts together pretty much as you want. Shall we add a foot to the length of the living room? No problem. A small addition to add a Jacuzzi in the left corner? Presto. You'd like the tea room on the other side of the house. Just press the button and the computer will make necessary adjustments in the plans and materials needed.

When the plans are finished, they are sent electronically to the factory, where they are cut from an assembly line almost one-third of a mile long. It will take less than one day for a crane and seven workers to get the roof and walls in place, and thirty to sixty days more for the finishing touches. In a two-story, three-bedroom house built this way just outside Tokyo, this included an electronic panel to warn you if the gas is leaking or the tub overflowing, and a dumbwaiter to pull drinks from the kitchen to the master bedroom at the touch of a button. The house also included a deck and small greenhouse. Cost: $110,000 plus the cost of the land.

The reaction from a group of visiting American home builders was mixed. One said, "You'd never guess it was factory made unless someone told you." Others said that the markets in the two countries were so totally different that mass customized housing would never work in the United States. It seems that we've heard that before about a lot of our once-strong industries.

One of the differences, for example, is that the Japanese use a wall material, called pelk, that looks like adobe. It doesn't burn, provides better insulation than concrete, weighs a quarter as much, can be molded into any shape, and doesn't have to be painted. Will the $6 billion U.S. housing industry adopt this material and other processes from the Japanese? Will American housing become mass customized?

"Most U.S. contractors are really assemblers who build houses using ready-made materials and equipment," says Tadashi Konishi, manager of Misawa Homes's International Product Department. "The problem with being an assembler, however, is that customers have a hard time distinguishing you from the next company. The smaller builders in particular, would like to create an identity by bringing in Japanese technology and equipment."[4] The largest prefab manufacturer in the U.S. turns out 2,000 units per year, a major Japanese manufacturer can produce 40,000 units. If the

U.S. home builders don't turn to mass customizing, they may end up out of business before they ever see 2001.

Once mass customizing is technologically feasible, it takes place in all sectors of an economy, including agriculture. I was lecturing about mass customizing in Italy a few years ago, wondering if the concept made sense there, when one manager raised his hand and said that he was in the fertilizer business. When Italy had an agrarian economy, he said, natural animal manure was used for fertilizer. During Italy's industrial economy, he continued, farmers used bulk fertilizer. "But today," he concluded, "we custom blend fertilizer, according to a number of variables such as soil, slope and sun, for each hectare of farm land. Some day we will customize the blend for each square meter, right as it is mixed into the earth."

The mass customizing of products can occur at various points along the line from design and fabrication to sale and delivery. This has been going on for quite a while, but generally without having been viewed as such. Modularization is one example. The parts are standardized, and the final product is customized at the end of the chain. This is point-of-sale customizing. The customer buys one manufacturer's turntable, another's amplifier, and speakers from a third. The final stereo system is tailored to one individual's specific needs, though the parts are not. The stereo can be placed on a modular set of furniture, and so on. You can buy a plain vanilla couch, for example, from the Swedish furniture company Ikea Svenska A. B., and then choose from a wide variety of fabrics and pillow styles. The result is a customized purchase for the cost of a mass-produced item. There is a very important lesson here, and one company captured its essence succinctly: "*Every buy is customized, every sale is standardized.*"

The core product of the industrial economy, the automobile, is treated very much this way. In the early days of the assembly line, the auto was a very standardized affair. After the introduction of the annual model change in the 1920s, product variations began to creep in more and more. To take the classic example, once different colors were available, customers began to choose the features they wanted. As the number of choices and options accelerated, fewer and fewer cars produced had identical mates. Of the first 850,000 Rabbits manufactured, less than 15,000 had an identical twin. All the others had some or several variations, giving them each a mass-produced uniqueness.

Although the decision to create any particular combination of characteristics lies with the manufacturer in most cases, the customer can also make the determination. In essence, customers can walk into any automobile showroom and order their own customized car that is mass produced and mass delivered. There is virtually no extra cost, and the waiting period is generally a matter of weeks, and the Japanese are working to get it down to a few days. (From the car company's perspective, the mass customizing is done by the dealers who place the order with the factory. The dealers' orders are a combination of individual customer orders and the dealers' inventory based on local experience. The decision as to what cars to produce is therefore made at the end of the chain, in the delivery system.)

In addition to the examples above from agriculture, construction, and industry, mass customizing will also be found in the service sector. During the industrial economy, for example, people used road maps to find their way around. They were mass produced, and may or may not have had the detail wanted. Today, however, road maps may soon become collectors' items, as computers start to replace them and plot the paths of tourists, salespeople, and fire trucks.

Logistics Systems Inc., for example, is negotiating with hotel chains and auto clubs to install a point-to-point map directory, which for a few cents each will route travelers to their next destination. The made-to-order maps will include such information as restaurants and places of interest along the way. The product is the intangible database, run as a customer service on the corporations' computers.

According to the *Wall Street Journal*,[5]

> *all three U.S. auto makers, as well as Japanese and European competitors, are working on even more advanced systems that will be installed right in their cars. Chrysler Corp.'s prototype, for example, has 13,000 maps of various sections of the country programmed into the car's computer. The device tracks the car's progress across the maps as they flash on a dashboard-mounted screen, using signals beamed from the government's Navstar satellite system. The auto makers believe their systems will be ready for market by 1990.*

Smaller companies plan to market computerized dashboard maps even sooner. They will pick niches for their services, such as package-delivery services, trucking companies, salespeople, and emergency vehicles.

For years, television was dominated by three major networks. Public television and some local unaffiliated stations reached only a small portion of viewers. Cable television, however, signaled the end of general programming for undifferentiated, mass audiences, and the beginning of mass customized programming. With the proliferation of channels, viewers have an enormously wide choice; still not a distinct program for each individual viewer, but often more variety to choose from than necessary. This is semicustomized viewing, with programs addressing narrow and highly specific customer groups. Cable combined with videocasette recorders, transforming scale plus time, creates something closer to the final customized view.

Rocky Aoki, founder of Benihana of Tokyo restaurants, has started a seafood chain that utilizes the new concept. Serving an average of fifteen different fresh fish daily, customers can have their choice stir-fried (Cantonese, Szechwan, Shanghai, or Thai style), broiled (California, Japanese, or French style), steamed or poached (Chinese or Norwegian style), sauteed (Grecian, Parisienne, Florida Indian River, or Louisiana blackened style), char-grilled (Key West, Nouvelle Cuisine, or Roman style), or deep-fried (Tempura, Cajun, or Italian style). The result, with an average inventory, an assembly-line kitchen, and an easy-to-read menu, is 285 different dinners to choose from – about one for every seat in the restaurant. Since fish consumption is growing dramatically in the U.S. (from 10.9 pounds per capita in 1966 to 14.5 in 1985, and it may approach 30 by 1990), Kroger's supermarkets now print up to 150 fish recipes on the price stickers. The customer buys a standard fish and chooses a customized recipe at the same time.

Niches, as we will discuss in the section on mass customizing of markets, allow businesses to target their services to specific customers. In advertising, for example, if two people in the same town are watching the same television program but receiving different commercials, the producer service (advertising) is being mass customized. And if two, then why not twenty-two? Is this possible? Yes. This precise targeting is possible with pay-per-program television technology.

The customer may also self-select. Newspapers have

long used zoned advertising and regional editions. Customization is already done, right down to the bundle of papers for a single route; tradition prevents further customization, down to the individual paper. The technology exists for subscribers to select from a menu of syndicated columnists, cartoonists, and sections. Say, for example, you want Dear Abby but not Dr. Ruth, Garfield but not Doonesbury, the local leisure/lifestyles section and the op/ed pages from your local paper, the *New York Times*, and the *Wall Street Journal*, and targeted advertising. The distributor negotiates novel yet simple-to-calculate agreements with various syndicates. Think of the pricing opportunities! Wouldn't you pay at least double for such a newspaper?

In the new economy, information is the critical resource, and the floppy disk is the vehicle for using this resource to mass customize an endless number of products and services. The floppy itself is already a mass-produced commodity; the information any one floppy contains is what customizes it. And the customers have performed the final manufacturing and tailoring themselves. An example of mass customizing a service this way may be found in the greeting card. Here are the steps for making and sending your own greeting card, through the mail or electronically:

Equipment:
Personal Computer
Colorgraphic monitor
Color printer
Software
Communication line

Process:
Sign-on system
Call-up card program

Call-up purpose selection – Xmas, Birthday, etc.
Select type – humorous, poetic, etc.
Select cover – various forms
                   color type
                   title – Mom, Son, etc.
Select verse and signature
Transmit.

Receive:
Sign-on system, read mail, you have a card
Request card, see on monitor
Insert color paper and press to print
Store, to see in future
Forward, transfer to family members system.

Alternate:
Send via MCI paper
They will store system and mail for about $2.00
Mail to distribution list
Personalize, add your own statement
Draw your own card.

Critic Ellen Goodman thinks that this mass-produced uniqueness is "the thoroughly modern illusion." Indeed, many of the applications do deal with trivia and a false sense of personalization. However, they also involve tens of millions of dollars. When Burger King advertised "Have it your way" they were trying to differentiate their commodity hamburger from the competition's by telling you that you can customize your order. There is a lesson here, however, that transcends the pickle on the bun. How do you customize a commodity? The answer is that you *standardize the commodity, and customize the services that surround it.*

Historically, only the haves of the world could afford customized goods, while the have-nots had to content themselves with the cookie-cutter standard ones. Painting pictures by numbers is an example of mass customizing that tried to bridge this gap decades ago. Decades later, it is ironic that today's avant-garde artist (and the average artist of 2001?) specializes in computer graphics art that requires more mathematics than artistry.

Alvin Toffler believes that the computerized assembly line can bring customized products within the reach of the average person; the economic fruits of democracy raised to the highest common denominator rather than to the lowest – and for houses and cars, as well as for clothing, hamburgers, and birthday cards.

If given the choice between a standardized and a customized product at the same price, which would you choose? Once again, the practical message for business is fairly apparent: companies that can apply technology and marketing to produce mass customized goods gain an advantage over competitors that cannot.

Customizing the mass market, in addition to the mass product, is another route that businesses will take increasingly in the future. Early markets were local affairs, defined in basically geographic terms. The beginnings of any one local market were probably rather simple, carrying a limited array of fairly undifferentiated goods. As markets in an agricultural economy grew, they became more sophisticated, differentiating the products they offered and appealing to more sectors of the local population. Major marketplaces were stops on important trade routes, and another market segment became the export trade.

Industrial markets followed the same basic evolution, from undifferentiated mass markets to increasing differentiation of the market parts. It was not until very late in the industrial economy that the marketing function formalized this differentiation into parts with the name "segmentation" – any of the parts into which something whole can be divided.

Once a market is segmented, advertising can be targeted. Very few advertisers have to reach all the people who watch network television; they just have to pay for doing so. Companies selling nail polish reach millions of men, for example, aftershave ads reach millions of women, and cat food ads are seen by millions of dog owners. Whittle Communications has averaged 36 percent annual growth rates for a decade by placing its clients' advertising so that it only reaches potential customers.

The most simple of businesses today can generally identify different segments of its relevant market. Even in the most complex business corporations, however, segmentation schemes are limited. Most methods are limited to a division of the whole into only a few parts. Table 2 is a list of typical segmentation schemes used in business, and the typical basis, number, and elements of each.

Table 2

Market Segments

Basis	Number	Elements
sex	two	male, female
income	three	lower, middle, upper
geography	four	north, south, east, west
age	four	children, teens, adults, elderly
occupation	four	blue collar, clerical, managerial, professional

There will, of course, be differences in each specific business. The point is, however, that businesses that segment their markets will create, at most, only a handful of different segments.

Segmentation is not a very refined way to differentiate parts of a market. So a more refined scheme evolved, one that uses an apt term, "niches" – any of the parts into which a segment can be divided. Upper-middle income, the Northeast region, toddlers, and doctors each represent a market niche in any one of the segments listed in Table 2. The means for differentiating are often mingled together, as in young professionals, working-class men, and female managers in New England.

By the time markets have been segmented, and then niches have been carved out, the number of differentiated parts is considerable, and few companies will serve every niche in every segment. Although segmented and niched, however, the market has not yet quite reached

Figure 3

Market Development

Local   Mass   Segmented
markets   ——→   market   ——→   market   ——————

its ultimate logic of internal differentiation. What is the final step, the unitary building block for the market whole in the new economy?

It is the "individual" customer. Units of one, whether a consumer or a corporation. But these are not the single consumers and firms who were reached with customized goods and services in the limited, preindustrial markets. Rather, in the same way that segments and niches are reached on a mass basis, individuals can now be reached on a basis that is simultaneously mass and customized.

Mass customization of markets means that the same large number of customers can be reached as in the mass markets of the industrial economy, and simultaneously they can be treated individually as in the customized markets of pre-industrial economies. The progressive development of the market can be visualized in Figure 3.

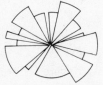

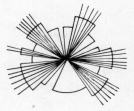

Market
niches

Mass-customized
markets

This evolution in the marketplace is not limited to national markets. What has not been clearly understood is that the same cycle also repeats itself in the global arena. The 1960s and 1970s were the decades of the multinational enterprise, which, for better or for worse at the time, was thought to be the most articulated form. In the 1980s the multinational corporation has come under attack for being inefficient, on the grounds that it caters to different national needs at unnecessarily high costs. It was never truly multinational, argue the critics; it was merely multidomestic, conducting separate businesses in several national markets at the same time.

The current winner, counterpoint to the multinational, is the global corporation. The global corporation operates on the assumption that the world is a whole, and therefore its inhabitants seek to make and sell the same thing everywhere. Strategic decisions about each business function (research, design, engineering, purchasing, manufacturing, marketing, selling, distribution, service) are made on a worldwide basis. All activities are coordinated across countries, and each individual activity is concentrated in the most advantageous country. The faith is that technology "drives the world toward a converging commonality," and the result is "the emergence of global markets for standardized consumer products."[6]

The distinction between multinational and global, however, is being put forth in a context that is both immemorial and inappropriate: The debate between uniqueness and universality is immemorial; to expect that one is correct and the other wrong is inappropriate. In reality, the multinational corporation and the global corporation are stages in business development, and not the last stages either.

Making and selling the same products the same way everywhere in the world is hardly a new idea, as Coca-

Cola, Hollywood movies, Levi jeans, and Sony televisions can attest to. More recent additions include the Uzi submachine gun, Chivas Regal scotch, and the world car concept. But the globalization fans have added a few new wrinkles. For one, they claim that "everywhere everything gets more and more like everything else as the world's preference structure is relentlessly homogenized. . . . Nothing is exempt."[7] As customers' tastes and demands homogenize, producers opt for the optimum singularity on every conceivable dimension: a globally standardized product or service at a globally standardized price, quality, reliability, and delivery.

The action of a single major player is what usually turns an industry to global competition. This occurs for any of several reasons, some of which include: to sell more than one's domestic market and leading export market can absorb; to utilize cheap labor and raw materials; to amortize high technological costs over several national markets; to respond to falling trade barriers or transportation costs. Whatever the impetus for the move, the result is always greater simplicity and higher efficiencies.

Globalization, therefore, overcomes one traditional contradiction of industrial economies. Until recently, mass market customers and those served as highly differentiated individuals were thought of as occupying the opposite ends of the economic spectrum. The one required low costs, the other demanded high quality, and the two were considered incompatible. Because of simplicity and efficiency, globalizers can be high quality *and* low cost simultaneously. The penalty, say the globalizers, is that they cannot offer a broad product line as well. The multinationalizers, for their part, argue that you can't get rid of enough cultural differences to standardize the offering to the point where global scale economies will make the difference.

Neither the multinationalizer nor the globalizer can overcome the other contradiction: being global and customized at the same time. As we saw with the nine dots in Aftermath, any contradiction can be resolved if viewed within a large enough context. The debate over multinational versus global corporations has been cast in the wrong framework. It is discussed in either-or terms, as to who is right:

- those who see uniqueness and argue that technology cannot homogenize cultural differences

versus

- those who see universality and argue that global standardization offers high quality at low costs.

Figure 4

Global Market Development

Multi-domestic → Global mass → Global
markets   market   segments

In truth, it is not necessary to take sides. These are not opposites, they are stages in the ultimate evolution toward global mass customization. Whereas the "ultimate" is a long way off, it is much more practical to see what is coming next. To do this, we must see the evolution within a redefined context. Here is what has and will happen.

The progression of market development that occurred on a national scale will be replicated on a global scale (see Figure 4). In the same way that multiple local markets yielded to an undifferentiated mass market on a national scale, multiple national markets will yield to a mass market on a global scale. Whether national or global, the undifferentiated mass economy brings enormous benefits of economies of scale in production, distribution, marketing, and management. The vehicle in the industrial economy is the national corporation, and

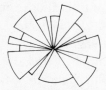

Global
niches

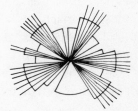

Global customized
markets

in the new economy it is the global corporation. But the global corporation will not stop at mass standardization any more than the national corporation stopped there.

At the beginning of the mass marketing era in the United States there were plenty of commentators who warned that customers would never want mass-produced articles that had no tailoring. Then Ford introduced the assembly line, and people accepted standardized goods because the benefits were brought within the economic reach of so many. It didn't take very long before customers opted for differentiation and producers responded by carving out segments they could tailor to specifically and still serve economically. The same thing will happen globally, and after a phase of global standardization, where possible, global market segments will emerge. These segments will cross national boundaries, and will gravitate toward high quality and low costs.

In the more distant future, it is not unlikely that there will be global niches and ultimately global mass customization. Though unlikely to occur in all businesses, it may occur in some by 2001. For a variety of reasons, some businesses are likely to stay national longer, and others are more inclined to fairly rapid globalization. From least- to most-likely candidates, for example, one might rank: cement, metal containers, packaged foods, industrial machines, drugs, telecommunications, automobiles, consumer electronics, semiconductors, computers, commercial aircraft, and jet engines.

Coca-Cola is used as the clear example of global standardization, yet in the Cola segment alone they have staked out new, classic, diet, and caffeine-free niches. When there was only one Coke, it was the same worldwide, now each niche will be the same worldwide. When differentiation comes, it is from market segmentation, not from national geography. When this is not the case, it is because organization politics has extended over the debate. In a sense, both points of view – the unique and the universal – are correct. There is differentiation between segments and between niches; there is global standardization within the mass, within the segments, and within the niches.

The progression described above does not necessarily portray historical stages of evolution in a strictly linear unfolding. For some companies, these stages represent increasing conceptual complexity, and may overlap or skip in time. For example, a company may jump from domestic market niches to global market niches. Also, those at the top end of a market may move to global market niches because of lifestyle and taste homogenization, whereas those at the bottom end may move to global niches because of cost reductions; and both ends may move to global niches before the middle-market shifts.

Whatever the particulars, the object for a company is to understand this progression, determine where in the progression each of its businesses stands, and map out a particular approach for each according to its position in its particular global life cycle.

The ultimate logic of ever-finer differentiation of the market is markets of one; that is, meeting the tailored needs of individual customers and doing so on a mass basis. Again, computers are making this more and more of a reality. An example of this trend is found in the way that mass retailers are using transactional terminals to customize their offerings.

Customers have become used to seeing terminals in retail stores, but it is the salesperson who usually deals with them, entering and getting back information. Transactional terminals, however, conduct business directly with a customer, helping to close a sale. Merchandisers are experimenting with these machines in the mass-customized marketing of everything from paint and hair coloring to shoes, clothes, and eyeglasses.

Whereas a mass-customized product is a one-of-a-kind manufacture on a large scale, the mass-customized market takes products of standardized manufacture and locates the one particular selection that is tailored to the individual's need. Each *customer's individualized needs are satisfied with mass-produced goods*, and the *customizing occurs at instantaneous speed in the matching-up process*.

Paints, for example, were rather variable in earlier times. Economies of scale called for standardized paint production in early industrial times. Henry Ford's famous remark, "You can have it in any color, as long as it's black," epitomizes this standardization. And the only reason the cars were all black, instead of pink, is because black was the fastest drying paint. When annual model changes came to the industry during the last quarter of the industrial period, customers could order a wide variety of colors. Shortly after that, people could also buy custom-blended paint in the hardware store. This customization occurred because the final stage of customized manufacture was moved closer to the space of the customer – in the hardware store. The customer selected a color from a color sample sheet, the clerk added specified drops of different pigments to a basic stock and mixed them for a few minutes to produce the customized product. Industry leader Benjamin Moore & Co. has taken the process even further. They supply paint dealers with a computer that measures the light frequencies of a color sample and therefore allows for a mix to match perfectly. The result has been 20 percent more Moore sold in stores that have the machine.

Magic Mirror, by L. S. Ayres & Co. in the Midwest, is perhaps the most fantasy-like embodiment of mass customizing a market. Even women who love to shop are limited in the number of clothes they can try on by the amount of time it takes to get into and out of each outfit. Speeding things up by looking at pictures or holding up a hanger doesn't do the trick. The solution is an electronic dressing room called Magic Mirror, where the customer sees a reflection of her own face, hands, and feet, and the computer creates the clothed rest of the picture, shaping the figure to conform to her own, and changing outfits as fast as the customer wants to. Final choices are generally made by literally trying on the narrowed down choices. Ayres put the entire Liz Claiborne 1985 spring collection on the system in three stores and the brand's sales soared 700 percent in one week. The mirror was invented in France, and distributed in the U.S. by Fashion Systems Corp., who hoped to have 150 of them leased in 1986. Leasing, program, and promotion costs are shared between the retailer and the clothing manufacturer.

Other systems focus on specific parts of the body. Elizabeth, an electronics makeup system by Elizabeth Arden Inc., measures skin-tone, recommends, and displays several different makeup treatment portraits without ever applying anything to the face. Split screens suggest four different suggested treatments simultaneously. The electronic "applications" are instantaneous and they overcome a person's reluctance to being made up in the store. Arden ordered nineteen systems in 1985, at $40,000 each. For the other end of the body, Florsheim shoes uses the computer to expand their inventories. With it, they can show hard-to-fit customers about 250 styles that they can't afford to keep in stock, and order them from the central warehouse at the touch of a button.

What happens if you have 4,000 styles in your product line? Eyeglass retailer Cole National solves the problem by entering a customer's facial characteristics and favorite colors into its computer, thereby narrowing the choice to the few glasses a customer should try on. Cole expects to computerize some of its Eyeworks outlets in Sears and Montgomery Ward.

According to the Marketing Science Institute, because of current population trends, real growth of retail sales is expected to be a very slow 2.3 percent through 1990. General merchandise stores (department stores and chains) are estimated to grow 2.4 percent, building materials and hardware stores 1.9 percent, specialty apparel and accessories stores 1.7 percent, and food, liquor, and drug outlets .8 percent. Moreover, a Touche Ross & Co. study calculates that there is 50 percent more retail selling space in the United States than is actually needed. With this combination of slow growth and overcapacity, retailers will be looking to increase their market share in order to survive. Mass customizing, along the lines described above, is likely to be a major instrument in that effort. Again, another key to the future lies in using new technologies to deliver mass-produced goods and services to individuals on a tailorized basis and mass scale simultaneously.

When mass-customized technologies, products, services, and markets become commonplace, during the years between now and 2001, we can expect to see more organizations move in the same direction.

In an organization that is structured according to business functions, each part is just that – a part – and not a representation of the total organization. R&D, engineering, manufacturing, sales, finance, personnel, etc., are parts of a whole, like slices of a pie, and not the whole itself.

The Sloan model of organization at General Motors

that we saw in Aftermath is actually a major precursor of mass customized organization. In the Sloan model, planning and financial control are centralized, and operations are decentralized around a product division structure. Each product division is treated as a complete business (whole), yet it is simultaneously a part of the larger (whole) corporation. The structure within a division looks pretty much like the functional design where the parts are parts, and there is only one whole. In other words, *within* the division level of organization we enter back into the either / or world: An entity is either a part or a whole – functional departments are parts of the division whole – and there is none of this confusing and contradictory talk. At the division level and above, however, units can be treated as both whole businesses and parts of the whole company at the same time.

This paradox in organization design came to light only about fifteen years after scientists had learned to accept their own paradoxes – those of the physical world: quantum mechanics, relativity theory, and the Heisenberg uncertainty principle. Science, technology, and business all had to accept the fact that mutually contradictory phenomena do exist simultaneously, and that our models – scientific and organizational – have to reflect this strange truth.

The next major advance in the same direction was the Strategic Business Unit. The SBU was developed in General Electric during the 1950s. In the terms of our discussion, it moved the part-whole paradox down to what seemed like the smallest unit possible. The SBU was defined as the smallest identifiable whole within the larger whole. The criteria for determining the smallest whole within the larger one are: a defined set of *resources*, applied to meet a defined set of *market* needs, through a defined set of *products* and/or *services*, in relation to a defined set of *competitors*. Note that these criteria are for business units, and while they translate into organizational units, the two are not identities. Like the discoveries that the atom, then later the electron, and later still elementary particles, are not the smallest identifiable whole, the creation of the SBU has moved us closer to this same discovery in organizations.

A logical next step is to move the parts-whole simultaneity down to the level of the individual. In the same way that the ultimate logic of internal differentiation in *markets* is markets of one reached on a mass basis, the ultimate logic of internal differentiation in *organizations* is organizations of one united together in a larger whole. Only this time the individual in the organization is seen somewhat differently. Each individual *is* the organization. The individual is not just a part, a cog in a wheel. Rather, the individual is seen as a whole within the corporate whole; not in the manner of the human potential movement, as in "she's a whole person," but as a whole business and a whole organization at the level of individual. One expression of mass customizing an organization was the "intrapreneur."

In his successful book, *Intrapreneuring*, Gifford Pinchot III defined an intrapreneur as "any of the 'dreamers who do.' Those who take hands-on responsibility for creating innovation of any kind within an organization. The intrapreneur may be the creator or inventor but is always the dreamer who figures out how to turn an idea into a profitable reality." The entrepreneur, he contrasts, is "someone who fills the role of an intrapreneur outside the organization."[8]

The resurgence of entrepreneurship in the new economy is a major plus. Despite the large number of people starting their own businesses, however, most people still continue to be employed in large corporations. Finding ways to liberate the entrepreneurial spirit and behavior in that majority who remain employees is a very worthwhile endeavor.

Edson DeCastro created Data General because he couldn't get his previous employer, Digital Equipment, to support his idea for a revolutionary new computer. His company became a major competitor of his previous employer. Steve Jobs left Atari and teamed up with Steve Wozniak, who left Hewlett-Packard, for the same reasons. As almost everyone knows, they created Apple Computers. The point is that if innovators can bring their ideas to be profitable businesses within their then-current employers, most of them would probably do so. Because so many feel they can't, the venture-capitalists benefit more than the large corporations.

In an opening memo to CEOs, Pinchot states that the key to keeping these innovators in the big companies is to realize that they are "working almost entirely *for themselves* within the corporation." In our terms, they are wholes within wholes, both operating simultaneously. They are individuals who succeed in customizing their results to their own intentions, within the mass(ive) corporation, to the benefit of both. The intrapreneurs masscustomize their organizations.

Pinchot's emphasis is that the innovative corporation need not be an oxymoron. Small and large coexist, not side by side, but one within the other; and *both* can be creative and successful. The customized fulfillment of individuals' needs, and of innovative products and services, can occur on a mass scale inside large companies.

There is still something missing, however, and it has to do with individual economics. I taught at the Harvard Business School for eleven years, and all that time I thought that managers played a *money* game. They don't! Managers play a *power* game. Entrepreneurs play a money game. One way to get entrepreneurship into big corporations, therefore, is to change the rules of power. The other, complementary, path is to make it a money game.

Pinchot's solution is not a true money game. It is the positive side of power, not of wealth. Like the human relations school of thought in the 1930s, and the need achievement idea of the 1960s, this approach emphasizes that the psychology of success matters more than the riches. It offers a symbiotic relationship where the individual gets the psychic rewards and the company reaps the economic ones. Pinchot would have large companies create "intracapital," a form of scrip. It is "something akin to" ownership and capital, but not the thing itself. Instead, intracapital comes in the form of freedom and security to do your own thing. This is an accumulation of psychological capital without the actual financial payout.

Another mass-customizing approach that has gained enormous popularity in the new economy has combined entrepreneurship with large corporations by making it a real money-owner game. This is franchising.

The modern form of franchising began in the United States at the turn of the century in automobiles and soft drinks, followed in the 1930s when the petroleum companies franchised gas stations and wholesalers first began to franchise their retail outlets. The use of fast food and other brand name franchising didn't begin until the 1950s. The concept has been around for a long time, with roots stretching down to the middle of the industrial period. There are aspects to the new boom in franchising, however, that differentiate it from previous growth periods, and that give it a particular role in the evolving mass customization of organizations. To appreciate this role, let us look briefly first at what franchising is, what forms it takes, how extensive it has become, and what makes for a successful franchise.

The origin of the word is from a granting of freedom or privilege, which is relevant to the spirit of entrepreneurship today. Franchising is a form of marketing, a method of organizing a distributive service, in which a parent company grants an individual or a relatively small company the right to do business in a prescribed manner, over a certain period of time, in a specified place. The franchisor also provides the franchisees with necessary central services, such as advertising, pricing, inventory, and quality controls. In return, the central business is paid a fee, such as a percent of sales. There is another way to look at franchising: It is the exchange of the information and technology of the center for the financial risk at the periphery.

Types of franchise systems vary, depending on the levels of the distribution chain (manufacturer, wholesaler, retailer, outlet) which are linked by the arrangement. Auto dealerships and gas stations, among the first franchise arrangements, link the manufacturer and retailer. Soft drink bottlers link the manufacturer and wholesaler. Auto aftermarket stores and hardware and drugstore chains hook up wholesaler and retailer. Others franchise a trade name as a licensor retailer. The latter may also sell the rights to a territory, which is then divided and subdivided down to specific outlets. This looks rather like an organization pyramid, but with individual ownership at every level.

During the industrial economy, franchisors seemed to be large and have money, whereas the franchisees were often the holders of small businesses with marginal resources, looking to hitch their wagons to some stars. Since the 1950s, the "little person" is as likely to be the franchisor as the franchisee. Successful fast-food chains, in particular, often started this way. The owner of a successful restaurant helps a friend or relative start and run a carbon copy. The process is duplicated until the originator starts selling franchises.

Franchise experts generally agree that the principal ingredients for a successful franchise are: a sound and easy-to-replicate concept, a proven prototype, adequate financing, and a good relationship between franchisor and the franchisees.[9] With these criteria, it seems as though almost anything or any non-thing can be franchised, particularly if it involves the retailer who licenses a trademark or trade name. Thus, franchises have included: accounting, credit and collection services, employment agencies, printing and copy centers, tax preparation, home cleaning, modeling agencies, convenience groceries, fast foods, hotels and motels, laundry and dry cleaning, travel, dating services, furniture refinishing, funerals, gift wrapping, personal shoppers, babysitting, lawn care, skin care, haircuts, cookies, and roommates. Even banks are getting into the act.

The manufacturer-retailer franchise accounts for around 80 percent of sales through franchise chains, although it is through a comparatively small number of outlets. The trade name retailer, on the other hand, has been the fastest growing type, accounting for around a third of all franchise units; yet this still represents only about a tenth of all franchise sales. Franchised businesses, in general, now account for one-third of retail sales in the United States, and the Commerce Department estimates this will be one-half by 2001. Even more impressive is that, while 90 percent of all new businesses fail according to the Department of Commerce, the annual failure rate for franchises is less than 4 percent. Considering the number of inexperienced people who go into franchising, this is truly remarkable.

With its product focus, the industrial economy touted, "Build a better mousetrap, and the world will beat a path to your door." With its market focus, by contrast, the new economy seems to be saying, "Build a better path, and the world will buy your mousetraps (or mouseburgers)." The American Dream gets realized and replicated through organization and marketing, as much as through the products and services offered.

A desire commonly expressed by management is to reinstate the entrepreneurial spirit into the large corporation. Franchising may be an opportune and (under-appreciated) way to do this. Unlike the intrapreneur, the franchisee is, significantly, a true owner.

The franchise owner's statement, "it may not be much, but it's mine," refers to his or her business as much as to anything else he or she may own. Earlier, we saw how the Japanese are mass-customizing new house construction. Americans are taking already-built housing and, through organizational rearrangements, are mass customizing as well. Houses are individualized at least in comparison to undifferentiated apartment houses in which living space is rather uniform. Moreover, the custom-variety house is generally expensive and owner-occupied, whereas a comparable apartment building unit is generally cheaper and rented. How do you get the advantages of individualized ownership and a large apartment building simultaneously?

The answer, of course, is the condominium. Each unit is both a whole and a part of a larger whole, at the same time. The outer shell of the apartment is usually the same as all others in its vertical row, but the occupant-owner can tear down walls and make other customized alterations without a landlord's okay.

Franchising is a way of organizing the distribution part of marketing. It is also a way to mass-customize organizations, by offering entrepreneurial freedom and effective central control simultaneously. It is exemplary of a holistic conception of organization, wherein each part is simultaneously the embodiment of the whole. It is also an example of how, in the new economy, the core function moves downstream, from production to marketing. Other new forms of organization also are likely to evolve in the downstream end. Retail services, whose central purpose is located in the end of the distribution chain, closest to the customer, is a likely place for innovation in organization. Thus, it is not surprising to see mass customizing in stores; moving, for example, from small neighborhood outlets specializing in meats, produce, and packaged goods to supermarkets that combine everything in one large space.

As we saw in the section on markets, small entities joined together into a mass, with the characteristic loss of individualized attention. Competitors of large supermarkets found niches that the giants were not filling. Specialty stores, convenience stores, gourmet shops, and the like came into being. How did the supermarket firms respond? By mass customizing, niching spaces within the building itself. Enter a supermarket today and you are likely to find the gourmet corner, the delicatessen next to the meats, the bakery next to the packaged breads, the salad bar next to the vegetable section, the generic goods in plain wrappers next to the branded items, and the bulk goods in barrels, like the old-time grocery store. Supermarkets are both small and large, both specialized and all-purpose, simultaneously.

The same thing happened to clothing, furniture, and hardware. They each had existed as individual shops, until they were combined into the same space in department stores. Through the decades these grew larger and larger, and as they did they lost their quality of differentiation – one store from the other, and one department from the other within the store. Store decorators tried to institute themes to remedy the sameness but could do so only to a limited degree. The solution came in the form of increased internal differentiation without loss of size. Boutiques opened up within department stores. Specialty corners set up next to standard departments. Today's department store is a honeycomb of very distinctive niches, all massed together yet different at the same time. Some, like Macy's in Herald Square and Fortnum & Mason, combine the upscale supermarket with the department store.

The stores organized their shopping space along the same lines. Small, specialized stores centered themselves around a village square. As suburbs grew, the large supermarkets and department stores moved into the shopping centers. The next step was a logical one – to combine the small and the large simultaneously in the mall concept. Malls are at once both large and small. Customers move through common open space, not buildings and doors, to go from one store to the next. The mall embraces a complex totality within a single expanse. It is a single, internally differentiated mass, with each part catering to the needs of the suburban villager.

When dealing with the fundamental transformation of an economy, it is essential to grasp the abstractions on which it is premised. However removed these may appear to be, ultimately they derive from our understanding of the fundamentals. The basic abstraction of the new economy, which we have focused on in this chapter, is the simultaneous existence of mutually contradictory phenomena.

A clear drawback of the industrial-based paradigm, on which the notion of economies of scale is built, is that it requires us to operate under the constraints of an either/or bind. In this model, either goods and services are produced in small volumes, in which case they are customized but have high unit costs, or they are mass produced, in which case unit costs are brought way down but high volumes make customization impossible. In the economy of scale model, it is not possible to custom-tailor products and services and *simultaneously* have the voluminous, uninterrupted production runs that are necessary for low costs. You can have one or the other at the same time, but not both.

This is the same problem we encountered in the chapter on Any Place where we talked about the wave-particle controversy in science and the centralization-decentralization dilemma of organizations. Our models are still built around false dichotomies that say you can't have it both ways, wanting your cake and eating it, too. Here are some managerial either/ors:

Either	*Or*
centralized	decentralized
headquarters	field
staff	line
strategy	operations
planning	implementing
important	urgent
task-oriented	people-oriented
corporation	individual
cost	quality
process	structure
level	span
flexibility	order
specialization	integration

New models, therefore, have got to overcome this either/or dilemma and deal with the simultaneity of business opposites. The simultaneity condition says that we must accept the coexistence of mutually contradictory phenomena without trying to resolve the contradiction. In the either/or dichotomies of an industrial paradigm, this is not possible. As in the example at the beginning of the chapter, a shirt is either custom-tailored or mass-produced. We have seen, however, that the new technologies will permit customized manufacture on a mass basis. Rather than being limited by the paradox, they seem to embrace and transcend it.

The power of the hologram lies in the fact that, if the image is broken, *any part that remains will reconstruct the whole!* For this to happen, all information about the whole must be present somewhere in each part. This unique property is unlike the mechanical paradigm of the industrial economy, in which the whole is merely the sum of all the parts. Rather, all information about the whole exists in each and every one of its parts.

If the whole exists in every one of the parts, as well as in the sum of them, then what space does the whole occupy? If the whole is everywhere, it is, equally, nowhere. The whole has no space dimension to it. If this is always so, if it occurs all the time, then the whole has no time dimension, either. If the pattern of a holograph has no time or space dimension, it exists only in frequencies, whose transformations are experienced as objects. This is a bizarre world, not the one we experience regularly. Yet it exists, as part of the universe.

We know this to be true scientifically and technologically. Is it also true for business and for organizations? Currently, the answer has to be: only somewhat, but becoming more so. Remember, these are the last transformations to take place, and each one raises the paradox in a philosophical way. Until the seeming contradiction is accepted in our technologies, it will be difficult to embrace it in our businesses and organizations. And we are still grappling with the technological causes of the paradox.

Medical technology, for example, can keep a patient alive under extraordinary circumstances, creating ethical and legal debates as to when a person is truly dead. If the heart or brain is "dead," yet the body lives with mechanical support, is the person alive? Only parts of the person are alive, yet many take this to mean that the (whole) person therefore lives. When families permit donor organs of the deceased to be implanted in others, a part of their loved one goes on living. Moreover, every cell of that donated organ (part) contains the entire genetic foundation of the departed. Does the whole exist in the part?

This is a fundamental abstraction, manifested in science and technology. If we can grasp it, it will be a powerful key to our understanding of society, of business, and of organization, which ask the same sorts of questions.

Can it be said with equal possibility, for example, that the entire family resides in each member, that the entire army resides in each soldier, and that the entire corporation resides in each employee? Can it also be said with equal possibility that the entire corporation resides in each of its products, and in each of its services? Certainly, from the customers' point of view, this may be true.

Technologies are bringing us toward more complete paradigms with which to build our businesses and organizations. The shift from a mechanical to a holistic paradigm is occurring in science and in technology. It is logical, therefore, that it will move next into our constructs of business and of organization.

Hello from the Future

A man was looking on the ground for his lost key. When his neighbor asked him where he lost it, he pointed to a different place. "Then, why aren't you looking there?" asked the neighbor. The man replied, "Because the light is better here."

— *Nasrettin Hoca*

Much like the man in the Nasrettin story, we waste time when we use inappropriate models simply because they are available. The alternatives I have offered, however, are not just theoretical suggestions for a hypothetical and distant future. Today, holistic logic and transformations of time, space, and mass are affecting businesses, organizations, and the people in them.

The economic shift that is occurring today is producing an economy of ambiguity and paradox, called, variously, a "postindustrial" economy, a "service" economy, and an "information" economy, yet none of these descriptive terms does justice to the full complexity of what is happening. The advent of this new economy – however "postindustrial" – certainly does not mean the end of industry, any more than the Industrial Revolution meant the end of agriculture. Agriculture, service, and industry are merely sectors coexisting within the same complex and interactive system. The true shift in this or any economy occurs when we adopt a new, all-encompassing model. The Industrial Revolution occurred when we stopped using agrarian models to run the agrarian and industrial sectors. The shift that must occur now is to stop using mechanistic, industrial models to run today's complex agricultural-industrial-service economy.

Einstein said that new frameworks are like climbing a mountain – the larger view encompasses, rather than rejects, the earlier, more restricted view. What we face today is an agrarian sector in which we have farmers who sit at computer terminals and never go near a tractor. In the industrial sector the car companies have become major banks. General Motors now can raise money more cheaply by issuing its own commercial paper than through bank borrowing; and Ford finances car loans for customers directly, often making more money on the car financing than on the car manufacture. Banks in the new economy, meanwhile, are finding that they can be more profitable as brokers than as intermediaries. Acting as principals, through the creation of loans and other assets, they generate high capital requirements and low returns on equity. By documenting loans at the time of origination to be liquid and packaged for resale, they can move them off the books at a profit and free up capital for other uses. This is what has happened, for example, when you find that your bank sold your mortgage to another financial institution. The line between commercial and investment banking blurs and, as the Glass-Steagall Act crumbles, commercial relationship managers, corporate finance experts, and capital market traders become very interrelated jobs.

Defining businesses from the producers' point of view, as was done in the industrial economy, is simply no longer workable. One hallmark of the ambiguous, new economy is the need to define businesses in terms of customers' changing needs. Many service giants of the industrial period learned this lesson and, for this reason, are now undergoing the most massive transformations in their histories. AT&T, the most successful and articulated model of a *service* organization during the industrial economy, is now completely revamping its businesses in awareness of the new economic reality.

AT&T, as we knew it until recently, was the largest corporation in the industrial world. It employed over a million people and had assets of almost $150 billion. It was in the telephone business in the United States, and the simplicity of this purpose, captured in its six-word mission statement, provided AT&T's direction for over seventy years: "One System, One Policy, Universal Service." Today, in the new economy, AT&T is in the business of worldwide electronic movement and management of information. Even AT&T's symbol has changed from the familiar electric Bell ring to the electronically encircled globe. In the words of its chairman during the transition, Charles L. Brown, "Ma Bell doesn't live here any more." To make this business transformation AT&T had to become deregulated, which required their divesting the twenty-two operating companies, over three-quarters of their mega-assets.

Today AT&T is also struggling to instill a new managerial mindset. The uncertainty of the marketplace has replaced the relative certainty of government regulation and, accordingly, AT&T must evolve a risk orientation. Its strategic planning must involve competitive analysis for the first time; its capital recovery programs have to adapt to short life cycles; its costing and pricing methodologies must abandon cross-subsidization and national price averaging, and each product and service must be calculated on its own economics.

Employees who were nurtured on a culture of up-through-the-ranks, lifetime careers, status consciousness, intense loyalty, and consensus management are still grieving and groping with the organizational adjustments that followed. The adjustments happened because the technologies of the new economy launched businesses which were both eating into AT&T's monopoly and also holding out promises of growth, but only if management totally revised its model of organization, which it has not yet been able to do. Despite the radical transformation of AT&T's business, its organization is still better suited for what was than for what will be. Unfortunately, this is the case for many businesses.

When it happens, there is an inversion of the appropriate relationship between business and organization, and it breeds bureaucracy. Remember, business and organization are not the same thing: A business is the application of resources to create products and services to meet market needs, in relation to competitors; an organization is the way in which those resources get administered. Organization is the means to accomplish the business's ends. You have to know *what* you are going to do before you can know *how* to do it. Organizations arise, therefore, out of society's institutions (such as government, school, church, business, and so on), not vice versa. For managers, organizations exist to carry out business; businesses do not exist to carry out an organization.

My definition of a *bureaucracy: a business, or any other institution, that exists to carry out an organization.*

In fact, I will go so far as to state Davis's law of bureaucracy: Any company giving less than two-thirds of its energies to its business, and more than one-third of its energies to its organization.

Where this exists, it is not merely that the company's organization is lagging behind, nor that the organization-tail is wagging the business-dog. Something even more serious is operating: The mindset that began in the universe, and wound its way through science and technology to business and organization, has reached a dead end and collapsed. It is time for a new world view.

From the perspective of centuries, this new world view is only now blossoming. The older, mechanistic context is premised on a framework that developed from the early 1600s to the middle 1800s. The mechanistic approach attempted to understand the universe, the economy, a business, or an organization by analyzing its component parts at the expense of their interactions. Those who adopted this approach became specialists of their particular parts. All the parts, once sufficiently understood and put back together, constitute the whole.

This breaking down of complex phenomena into their smallest constituents shifted the very purpose of inquiry from meaning and revelation to prediction and control. Experience gave way to experiment, and the empirical approach replaced the intuitive one. In the mechanistic science and technology of the past, the heavens were a mechanism and the world was a huge machine. Understand how the heavenly machine functioned and you could design your own, more practical versions. These machines became the engines of the Industrial Revolution.

The same mechanistic assumptions that went into science and technology also entered into the frameworks for the business and organization of the time. Even today, the basic graph in every elementary economics textbook, showing supply-and-demand curves, is based on assumptions of Newtonian mechanics that the two curves will "gravitate" toward "equilibrium." New science, applied to new technologies, yielding new products and businesses, for which new organizations were built – and all based on mechanistic models.

The mechanistic view in business – dividing things up into their smallest parts, giving specialized attention to products, to the resources that create them, and to the markets that need them – gave very little attention to interactions. Organizationally, this led early on to functional structures in which purchasing specialists dealt with physical resources, treasurers and controllers handled financial resources, personnel treated human resources, manufacturing handled the product, and sales specialized in the market.

When scientists reduce any whole – whether a galaxy, planet, organism, cell, gene, or elementary particle – to its fundamental parts and then try to explain the whole in terms of the parts, they are unable to explain the relationships that bind the parts together in and to the whole. Similarly, when managers reduce a whole to its fundamental parts – whether the global economy, an industry, a corporation, an SBU (a strategic business unit), or a manager, in a specific time and space – and then try to explain the whole in terms of the parts, they run into the same problem. The model focuses attention in the wrong place, on the wrong time, and – wrongly – on "things."

Einstein also said that the model tells you what you can observe, and to observe and manage companies today we need correct models. With the mechanistic economic models prevalent today we still cannot predict interest rates, forecast recessions, or control inflation and employment simultaneously. Economic theories predict price, not value. Value is a social phenomenon, woven politically into the economic fabric, and unaccountable by mechanistic theories. Even the more modern economics has a mechanistic orientation. Keynesians cannot stimulate demand without increasing inflation, and supply-side economists cannot create prosperity without removing environmental constraints.

The competitive model widely used by economists today is still based on Adam Smith's theory which assumes a perfect market mechanism. In other words, economists look at perfect parts of an imperfect whole, and cannot tie them together satisfactorily, for purposes either of understanding or of control. Managerial models have not fared much better. We are still unable to predict success or failure in mergers and acquisitions, accurately forecast new product launches, unite quality and productive efficiency, or combine innovation and hierarchy.

In sum, we wouldn't use farm models to manage a factory economy, and we shouldn't use factory models to manage an office economy! We need the same transformation from mechanistic to holistic models in business and organization that has taken place in science and technology. Managers steeped in the Western tradition will not accept the "soft fuzzies" of an updated, 1960s-style human potential movement. Fortunately, we don't have to look to Eastern cultures or to humanistic extremes.

Consider the simple example of changing your television from channel 2 to channel 7. With a mechanical knob, you would have to move it from 2, through 3, 4, 5, and 6, to get to 7. Using today's electronic selectors, the change is instantaneous and directly from 2 to 7. With the new technology, each channel has direct access to every other channel.

The same difference potentially exists for organizations. In mechanistic-industrial models of organization, if department 2 wants to communicate with department 7, it has to go through the chain of command – departments 3, 4, 5, and 6. Using information-based networks, rather than authority-based hierarchies, however, departments 2 and 7 can communicate without bureaucracy, directly and instantaneously. The technology exists to make this possible. It is management's mindset that is mechanistic, lagging behind in a model from a world that no longer is.

In a holistic model, the whole is not the sum of the parts, but rather the sum of the parts and their interrelationships. It is the relationships between the parts that give them their significance. Leontief's input-ouput analysis in economics is an example of a holistic approach, taking a single input and tracing its impact on every part of the economic system. The way the inputs are integrated into the whole is even more important than the parts themselves. For example, by giving machines image sensors for eyes, robotics for arms and legs, and computers for brains, the interrelationships of these technologies give us artificial intelligence.

In the mechanistic model the irreducible element is the part, in the holistic model it is the whole. The whole is not merely the sum of the parts, nor can it be comprehended in an additive way, gradually. It has to be apprehended in the same way that it is composed, instantly.

Not only is the whole considered greater than, and different from, the sum of its parts, but the whole can also be interspersed in all its "parts." Arthur Koestler coined the word "holon" for something that has the characteristics of being both a part and a whole at the same time. As such, it has both an integrative tendency and an autonomous aspect.

Consider our genetic code. The code for our entire biology is in *every* one of our cells. The codes for hair do not exist only in hair cells, the codes for blood do not exist only in blood cells, and the codes for bone do not exist only in bone cells. Rather, the codes for hair, blood, bones, and every other part exist in every hair cell, blood cell, bone cell, and in all cells.

Likewise, the model of the brain that is evolving is far removed from the machinery of the Newtonian age. This transformation in our understanding is mirrored in the transformation of computers, from early and quite mechanistic "electronic brains" to today's highly sophisticated – and holistic – mainframes. Both transformations offer a precursive view of holistic models for both business and organization.

Computers are moving into the realm of how brains think, apprehending the whole at once and not as a sequence cf parts, much as we apprehend a pointillist painting as a whole rather than as an aggregation of dots. AT&T Bell Labs, and a host of start-ups (Hecht-Nielsen Neurocomputer, Synaptics, Neural Tech, Revelations Research, Nestor) are testing experimental computer chips that mimic the way neurons transmit messages from various parts of the body to the brain, through compact and extremely interconnected networks called electronic neurons.

Separate thoughts, images, feelings, and the like do not each reside in single neurons in the brain, but rather are distributed throughout areas of the brain and reside simultaneously in many parts. Whereas conventional computer circuits' signals are processed step by step, neural computers fan out across the whole network and all of the transistors process signals in parallel. This is important because it allows machines to synthesize knowledge from random data. Shown a few views of a particular face, the neural networks will instantly recognize that face from any other angle.

New mathematics that use parallel processing architectures are making possible precise descriptions of intuition, affects, attention, and intention. According to Stanford University neuropsychologist Karl Pribram, these mathematics "allow holistic descriptions to be as rigorously scientific and precise as any that have been used in physics, chemistry, and biology."[1]

What science shows is true in the universe, as in the biological example above, is not limited to science. If it exists in the universe, then it is equally valid for application to technology, to business, and to organization.

In business, a strategic business unit is both a business and a part of the same business simultaneously. From this perspective, then, the entire company is present in every product and in every employee.

By 2001 our familiarity and acceptance of holistic modeling will reach into the management and organization of business, the way it is already reaching into psychological processes in the brain.

Holistic orientations will slowly gain credibility in management, *first as metaphor and then literally*. Its managerial utility will only happen after its acceptance by science and technology produces a substantial number of business applications for the essentials of the framework to be understood and embraced. Today, nonetheless, we already can see that a holistic conception advances not only computer science and computer technology but also our understanding of the computer business. Information comes in four forms: *words, data, image,* and *sound*. Think of the products and businesses that were spawned in order to serve each of these forms. Words were captured by telex and typewriters and disseminated by printing and publishing, to name a few. Computers dominated data capture and processing. Images were handled by cameras, movies, television, copiers, and facsimiles, among others. Sound was handled by radio, telephone, and tape recorders.

Each of these businesses operated rather independently of one another for decades. Sooner or later, however, the lines began to blur. To take a simple and very early example, movies started with image and combined it with sound. Computers started with data processing and have spread into words, image, and sound as well. Today's new "industry" combines all four forms together. So, those who thought they were only in the computer business found themselves competing in the emerging new information processing business, with a lot of others to whom they were suddenly very interrelated.

The new blending has created a new whole whose parts include: *hardware, software*, the communications *pipeline* that carries the information, the *exchange* that hooks all the hardware into the pipeline, *services*, and the *network* that relates all the pieces together. Each of these parts were, and in some cases still are, wholes themselves. Some were thought of as parts in a linear chain; telephones, exchanges, and lines, for example, but all still processing voice communications. The future, however, lies with companies that understand that their relationship with this new whole depends on their connectedness to it, and their compatibility with all other parts of it.

The drive for connectivity in the computer business is comparable and a predecessor to what organizations will also have to learn to do, namely, put all parts of the whole in relationship with one another. In the industrial paradigm, when the whole organization is divided up into its relevant parts, the problem becomes coordinating and integrating them back together. The approach and solution for accomplishing connectivity in the computer business will provide precedence for doing the same thing in organizations. Linear telephone tag, for example, will be superseded by the more holistic electronic mail; and, in time, networking will supersede hierarchy.

In the industrial model, once a product is bought in a primary demand market, it creates a separateness. This is brand loyalty, in its best marketing form, where the customer won't buy other competitors' products. The notion of locking customers in is being applied to the primary demand market for intelligent desktop machines. There are around 36 million desks in America. By 1987 around half have some intelligent machine such as a CRT or a programmable phone on them, and all will have them there by the early 1990s. The race has been on to get as large a share of the primary demand market as possible. The belief is that if your machine is purchased first, then you will capture all subsequent purchases as well, including all services, all software, the next generation hardware, and so on. This orientation is based on the classical assumption of sequentialness, the first in the sequence determines the rest.

Where computers and their software are compatible, this sequential determinism seems very strong. From a holistic perspective, however, it is not the first in the sequence that wins. It is the first to create compatibility that reaps the rewards. These are the companies that have unique products or services that can interrelate with other unique products and services which were previously incompatible with each other. Technology is the enabling device, making possible the connectiveness, creating compatibility of previously incompatible parts to create new wholes. Organization, like technology and business, needs to refocus on the compatibility of parts, such that each may access any other part, and embody the whole, simultaneously. In the computer, this is called RAM (random access memory). In the organization, this does not yet exist; and won't until there is a conception to house it.

Compatibility stresses the connectedness of each part to all other parts, and the full functioning of a part only exists when it embodies the whole within it. The importance of a part lies in its ability to achieve greater functioning (utility) through its ability to link up (connect) all parts in the whole.

Mutual funds, an example created in the new economy, behave according to this principle. A fund's power is not in the integration or coordination of the companies whose stocks are involved, but in their compatibility with one another around the defined criteria upon which the stocks are selected for the fund. No one stock has any meaning except in relation to the entire fund.

New concepts for management are more likely to be accepted in new businesses, run by new leaders, who are building new organizations than they are to take hold in established businesses, with established organizations, and established people running them. In each company, management will have to work out its own responses, and those who grasp the importance of the shift that is occurring will have a decided advantage in the future.

The new concepts in this book are not just theoretical suggestions for a hypothetical and distant future. Today, holistic logic and transformations of time, space, and matter are affecting businesses, organizations, and the people in them. Federal Express Corporation offers us a stunning example.

It is the first company in U.S. business history to spot an unidentified need in the market and, then, to grow a business from zero to over a billion dollars in annual sales through internal growth alone, in less than a decade.

The business concept was the brainchild of Frederick W. Smith, founder, Chairman, and Chief Executive Officer. Originally written as a Yale term paper, it survived his years with the Marines in Vietnam and was launched in 1973 in his home town of Memphis. The idea was for a new freight distribution system for a new economy. Air freight was slow, unreliable, and treated as an add-on to passenger service. Smith saw that there was a giant need for an efficient service to move high-priority, time-sensitive packages and documents. Fulfilling that need meant building a totally integrated logistics network, differentiated both from the low end of air freight and from electronic facsimile transmission.

The result is a company that "specializes in the door-to-door express delivery of goods and information throughout the United States and abroad."[2] To accomplish this, said one magazine, means that Federal has to combine "the dispatching capabilities of a nationwide-sized taxi company, the sorting facilities of a postal service without the time margin, the logistics of an airline whose passengers can't get themselves on and off the flights, and the customer service of a bank whose customers can demand immediate confirmation of every transaction."[3]

Despite such difficulties, the business has emerged as one of the fastest growing and most competitive in the United States. Fred Smith, more than anyone, created the business and has built the model for running it. The model applies the concepts we have explored. He believes that "Federal manages time, space, and what you call no-matter, as resources that are as important to us as people, technology, and capital." Here's how they do it.

Federal focuses on time the way a record-breaking runner does; your competitors are only there to spur you on, your real competition is the clock. In 1981, a Federal study showed that the "industry" leader, Emery, had a 65-percent probability of getting a package to its recipient by the next day. UPS, their more serious threat, committed to a 3:00 P.M. next-day delivery in its 24-city network. Federal used time as a major ingredient to define the business they were in, and to differentiate themselves from the competitive threat.

The public still saw the air freight business as the seedy guy with the cigar, loading sacks of potatoes and heavy machinery. Federal was out to change the image, because they saw themselves differently. According to one public relations message, "UPS is the best in the business at what they do – moving low priority, consumer-oriented parcels where emergency is not a factor. But Federal, in turn, is best at what it does – movement of the most vital and time-sensitive parcels and documents."

The unidentified need in the market was the time-sensitive nature of items in the new economy. Computer parts and tapes, vital medicines, legal documents, checks, and other high-value articles were often needed on a "must-have" basis. Fred Smith defined a time-sensitive item this way: "One for which the consequence of failure to deliver within a specified period of time would far outweigh any consideration of reasonable rate comparisons. If you offer two services and one of them is 15, 20, 30, 40 percent higher on one hand and it gets the documents or items there on time, and a cheaper service whose reliability is not nearly so great, the shipper will always take, if it is time-sensitive, the more reliable method, rather than the cheaper method. This is the essence of the definition of time-sensitive."[4]

Federal's original priority commitment was to deliver before 12 noon the next business day. Defining the time-sensitive business their competitors would have to follow them into, they also set the industry standard as they keep whittling away at the clock. In 1982 they shortened their delivery commitment to 10:30 A.M. In 1984 they offered a money-back guarantee if the package arrived after 10:30 A.M., and in 1985 they announced they would do the same if they couldn't locate a customer's package within thirty minutes of an inquiry. Their goal is to be able to locate packages while the customer is on the line; and, for customers who are linked into them, to enable them to locate their item directly and immediately themselves.

Each of these actions manages time as a key resource. "It's really a matter of understanding time from the customer's perspective," says Carol Presley, senior vice-president for marketing, "and driving it down to zero." Zero means psychologically as well as temporally. She would like to guarantee pickup, for example, within a certain number of minutes of a call. "Even though it won't get the package to its destination any sooner," she says, "it will get the burden of responsibility off the customer's mind."

Strategically, as well as tactically, Federal treats time as a valuable resource which needs to be managed explicitly. It is at the heart of the many ways they define the business they are in: The early "We're a freight service with 550-mile per hour delivery trucks" is linked to reliability in their "Absolutely, positively overnight!" and to their more recent "Time-sensitive paper trail." It is at the core of all their programs and initiatives for better customer service. People, technology, and capital are the support resources; time is the paramount one. And the goal is to "drive it down to zero," the instantaneous pickup, transmission, and delivery of America's high-value goods and information.

There are two major ways in which Federal Express manages space. One is the extraordinary distribution system that provides door-to-door express delivery throughout the country, and now to Europe. They have created this one for the physical distribution of goods, and are building another for the electronic distribution of images. The second way is by "customer automation," moving many of the service components into the customers' hands and space.

The distribution system combines trucks and planes in a single company, something that was unheard of before Federal made it necessary. Packages enter the system at the periphery, through one of 10,000 drop boxes, 400 business centers, 400 walk-up counters, 700 ex-Fox photo kiosks, or 12,000 van couriers and 6,000 vanless couriers in large office buildings. An entry point into the Federal system is within fifteen minutes of 90 percent of the U.S. population. As soon as the packages are picked up, a scanner reads and registers the zip code on the shipping label. The scanner downloads the information into the network, and the tracking begins within two minutes. Then, the packages are loaded onto planes headed for the Hub.

The Hub is the heart of the physical distribution system. It is located at the Memphis airport, and three others are planned for Newark, Chicago, and Oakland. The Hub is a space-age production function in what is basically a service business. The first of sixty-three jets arrives by 11:00 P.M. and the last one leaves by 2:45 A.M. In between, about 700,000 packages from all over the United States are unloaded, sorted to their destinations, and reloaded. They go from the planes, back to the couriers, and to delivery in the periphery the next morning.

The packages may hold a single sheet of paper, hundreds of millions of dollars in checks for clearing, fresh flowers, Florida grapefruits, electronic equipment, and even an organ for transplant. They can weigh up to 150 pounds, and they move along on conveyor belts at 500 feet/minute. Three thousand people do the job, many of them college students. (Twelve hundred people a month, including "60 Minutes" and the U.S. Army, come just to tour the facility.) The energy level and accuracy rate is incredibly high; this is anything but the graveyard shift. During the first moments of my visit there, I felt like someone on a 007 secret base – on the side of the good guys.

In their quest for service, Federal is always coming up with new services for their sending, delivering, proof of delivery, receipts, billing, and ordering supplies. Federal's orientation to these information services is to provide as many as possible, as instantly as possible, and at minimal cost.

Federal expects its package deliveries to quadruple in five years to over two million pieces daily, delivered in the same few number of hours. Despite this growth, according to Ronny Ponder, SVP Systems and Automation, Federal Express will not have to key-in any more information than they do today. Satisfying these criteria means automating the tasks by placing the customers on the electronic network, then letting them make the inquiries and get the answers themselves. For example, today they're billing more packages from devices that are in the customers' organizations than they were carrying in the entire system five years ago. By 1990, 50 percent of the information entries will be done automatically, directly by the customers. "Most companies would consider that goal enough to break a company," says Ponder, "and that's just the way we want it."

On the trail of customer automation, they have placed a few thousand prototype machines in the customers' space. The machine combines a computer, scale, keyboard, scanner, screen, and phone into the space of a breadbox. In user-friendly manner, the machine, among other things, automatically fills out the invoice and shipping label, updates an internal report, rates the package cost, sorts for charge-back to specific departments, and will track and trace each package upon request. The machine is so cost efficient that Federal Express can put it into the hands of customers who process only five packages per day. "Once it's in, of course," beams Ponder, "they start using it more and more."

For the person shipping once or twice a week, they are planning a slimmed-down version, called the Cube 2, that fits in the base of a pen-and-pencil set. "Its channel devices add value to the customer and productivity savings to us. Customers do our work easier and quicker themselves. Usage goes up. Everybody wins."

Federal's tracking system is perhaps the most sophisticated example of managing space. Imagine being able to locate instantly every one of a million packages, anywhere in the system. They constantly check pairs of an increasing number of scan points to make certain that packages aren't missing or misdirected. The status of packages leaving the Hub for a specific destination, for example, are reconciled with those entering that destination from the Hub. This watches for thefts and losses, and "catches the smoking gun while it is still smoking." In terms of both service and systems, Federal Express consciously manages time and space as key resources with which to serve customers, improve efficiency, and out-do competition.

As Federal Express grew larger and faster than anyone else in the business, they began to ask where the limits were and what to do next to maintain a strategic edge. The answer came in recognizing the shift from tangibles to intangibles, but the recognition came too late. The failure, called ZapMail, was to be like a national Xerox machine working on AT&T phone lines. Here we have a case study within a case study, a failure within a larger success story. The strategic logic behind it was correct, but the application was much too late and was leapfrogged by competitive facsimile technology. Attentive to our larger thesis, however, both the winner and loser premised their strategies on managing no-matter.

Although there were limits to how fast packages can move across the country, now measured in hours, information could move a lot faster than that. Moreover, the demand for *rapid* transmission of *printed* information began to outpace the demand for rapid movement of freight. ZapMail, a system for the electronic transmission of high-quality document duplicates was created to meet that need.

So long as information was in digitized computer language, it could be sent anywhere over the existing telephone system. Federal's fatal mistake was failing to see electronic mail and facsimiles as competitors to ZapMail. Electronic mail was not thought of as a competitor because, unlike ZapMail, the information had to be manually encoded to send it, and because it was not user friendly the way paper is. In terms of product analogies, Federal mistakenly believed that facsimiles were to ZapMail what the telegraph was to the Bell telephone system, and what the mimeograph was to Xerox. They saw facsimile as an early generation of product with nowhere near the value features of ZapMail.

Early thinking about facsimiles, in fact, was around what you could do with the facs machines and not how you could use a system to do what you wanted. Even though machine and network have to work together, the system itself is independent of whichever ground-based communication equipment is used. Facsimile technology improved rapidly, and its purveyors understood the importance of linking it into a complete system. Canon's FaxPhone 10, for example, lets you hold a conversation while you send and receive documents in the same phone call over the same desktop machine. MCI Communications delivers facsimile, simultaneously with voice, data, and telex services on a global scale. It was a painful $190 million dollar write-off when Federal abandoned ZapMail in 1986.

In sum, Federal manages time and space as resources that are as important to them as the more traditional ones of people, capital, and technology. Their unexpected competition from facsimile succeeded better than they did in doing the same with no-matter. Federal has withdrawn from being a network-oriented mover of information, and returned to being the best jet-oriented mover of packages.

Federal has also started paying attention to the applied concept of mass customizing. According to Federal's chief operating officer, Jim Barksdale, they can't engineer a solution for every customer, yet they do need to give specialized attention. Therefore, they engineer the systems very finely so that each system can adjust to specific customer requests.

The application of the concept is perhaps most clear in their customer calling centers where they handle almost a million phone calls every week. That's about a phone call for every four packages shipped, and each one is personalized. For any of us who have called a telephone operator or any other customer service phone number, we know how exasperating the experience can be. Federal would like to see the call as an opportunity to provide service. One measurement of their service level, for example, is answering the phone within four rings; they rate the fifth ring as zero on their service level scoring.

Another aspect of service that is customized to massive numbers is the 800 number that people can call to get directions from wherever they are to the nearest drop-off place. Federal also works to personalize treatment, so that when a caller gives an account number, their entire file and history with Federal comes up on the operator's screen. This is reminiscent of some of the better hotels, where the operators greet you by name when you dial them on the phone.

Another example is in shipping and billing. Federal created a (2) × (2) of ("Where I Ship It") × ("Where I Pay It"), each being either centralized or distributed. For each of the four segments (e.g., centralized shipping + distributed billing) they develop different systems, and the systems then customize to the client's needs. Their ORBIT system produces a quarter-million billings every twenty-four hours. There are 3,400 Federal meters on the shipping docks of their largest customers, handling about 100,000 packages per night. "Five years ago," says Barksdale, "the whole company didn't do 100,000 per night, and we had one billing form with no individualized MIS."

Now, when the Federal meter scans the bar code on the package, the information is locked into both their system and the customer's system simultaneously. The customer tracks shipping and billing their way, and Federal tracks the same packages its way, at the same time. Each uses its own customized system which, at the same time, is integrated with the other. The focus is simultaneously on both the customers' needs, the provider's needs, and the relationship between them.

When it comes to the customer, Federal has truly managed time, space, and no-matter in astounding ways. They have also figured out ways to mass customize their services. These impressive moves, however, are all focused on the business. What about the organization?

Federal Express is, no doubt, an extremely well-managed organization. There is no way for them to have started from scratch and become a billion-dollar-plus corporation and the industry leader without significant organization competence. Despite this, the organization does not yet match the business when it comes to applying these same new dimensions.

By current notions of the well-run organization, Federal is as good as the best in many ways. Their pay scales are the best in Memphis and in the industry. They have several incentive and profit-sharing programs. They have promote-from-within, open-door, and no-layoff policies. Their perspective on unions is that they occur when management fails in its responsibilities to its employees. "Companies that get unions, deserve them, and they deserve the kind that they get," says James Perkins, senior vice-president for personnel. "If management manages properly, employees don't need or want collective bargaining." They are totally nonunion.

They are also moving away from part-time work. In early 1986, 21 percent of their labor force was part-time, and 3 percent were casuals. In March of that year, they made everyone who wanted to be, full-time, guaranteeing a 35-hour week. When they were part-time, they got paid less if they finished faster than expected. The full-time approach is more congruent with the goals of moving the goods as quickly as possible.

It is a good example of their slowly growing awareness that they need to manage time as skillfully for their employees as they do for their customers. Sixty percent of Federal's employees have direct customer contact, and all of them are primed to producing in real-time for those customers. Therefore, they have come to expect that what can be done for one can be done for the other.

In 1983–84, for example, Federal almost doubled their work force in eighteen months, and many of these employees wear uniforms. At that time, however, there was no concept of just-in-time inventory. If an employee is an odd size, it can take weeks to get the proper-fitting pieces. Until that happens, the person cannot work. "How is it," they logically ask, "that we can deliver uniforms for customers overnight, but can't deliver uniforms for our own employees in the same time frame?"

Federal is working to get the lag-time down to a couple of days.

Medical insurance claims took six weeks to process. Why not "absolutely, positively overnight?" They've got it down to a week, and are working toward two days. Perhaps more important, they know what it is that they don't know: how to run an organization with the same real-time no-lag expediency as they run their business. The question is felt by the leadership, and improvements are genuine, but clearly there is no equivalent sense of urgency.

Making certain the customer has adequate supplies, such as mailing envelopes, and that the internal mail gets delivered in the same real-time are two other concerns that employees ask about. There is a "sunset provision" philosophy that says the sun should not set without a customer complaint being acknowledged, if not resolved. Some managers would like the provision to apply equally to customers and employees. Currently, it is a question of the shoemaker's children.

"I'm very frustrated by the organization," says one senior manager. "We have a bureaucracy that owns the organization. It takes too much time fighting to get the approvals that I need. Particularly among the staff, the focus becomes getting through the bureaucracy, rather than solving the problem. I've got approval on dollars and people, for example, but it will take all year to get the titles and structure approved, and by then this year's budget approval window will be closed."

This kind of lament is familiar, even in the best of organizations. What is important to us, here, is the context: time is a key resource, and the goal in managing that resource is instantaneous real-time delivery, to customers and employees equally.

Having redefined the context to create a real-time organization, Federal has turned to information-processing technology as a key weapon in their skirmishes (it is not truly a war) with bureaucracy. PRISM, for example, is an automated personnel system designed to eliminate paperwork and to make changes in real-time. Anyone with the proper identification code can enter the system and make major changes, such as salary and bonus increases, electronically. Managers concur that the system has been a success, and 90 percent of the paperwork has been eliminated.

With regard to their organization, Federal Express is not atypical, and it is better than most corporations. What we see is that *technology* made possible a high-speed integrated logistics system for a consumer business. Federal put to work the world's fastest and built a real-time *business*. Doing so was an act of true creation, what the jargon would call proactive. The *organization* is well-run by conventional standards; yet when it comes to applying the new dimensions to this last link, the lag is still there. Management is reactive.

Federal is a success story of the new economy. Faster than most companies, its visionary leadership realized that treating time, space, and no-matter as resources gave them a definite competitive edge. To use such an approach requires a mindset that starts and ends with the whole, that doesn't treat parts in isolation or as incomplete pieces of larger puzzles, that emphasizes relationships, and that understands parts in their relationship to one another and to the whole. Even they, however, have been slow applying these transforming insights to the management of their organization.

It is fitting, here, to end with a theme from the beginning of the book. When managers manage the consequences of events that have already happened, their organizations are doomed always to be lagging behind the needs of their business. The history of the industrial economy is strewn with companies that, once bright new lights, have since faded into obscurity, senescence, or death because their leaders managed in the past imperfect tense – that is, forever managing aftermath. They became bureaucracies: businesses that existed to run organizations. The way out of this dead end is in finding new models for managing in the new economy. Federal is one of the companies that has done so for its business, and it became a bright new light. We are awaiting the ones to do it for their organization, managing the before-math, in the future perfect.

Beforemath

"It is the theory which describes what we can observe."

—*Albert Einstein*

Many years ago I asked an executive responsible for the future development of a very large corporation, "What do you worry about most on your job?" His answer was startling. "I worry most about what my people don't know that they don't know. What they know that they don't know, they're able to work on and find the answers to. But they can't do that if they don't know that they don't know." During the past few years, many people have come to feel like this executive.

What does it mean not to know that they don't know? In some ways, this is the condition, "ignorance is bliss." It is only when a problem is identified and defined that people can go to work resolving it. Before that time, there is no problem. The same events may have occurred long before the identification of the phenomenon as an issue, but they are meaningful only because meaning has been attached to them.

The way in which people perceive a problem, a question, or an event determines what they will be able to know about it. The newborn, for example, is not able to distinguish itself from its environment; it must sense the environment as "not me" before it can develop any distinct sense of "me." The infant moves from not knowing that it doesn't know, to knowing that it doesn't know what is "out there" beyond itself, to knowing. Initially, this is frightening and confusing.

For all of us, there is both terror and exhilaration in

being on the existential edge. The shift in our view of the world from a mechanistic to a holistic perspective is like living on that edge. Distinguishing between a business and an organization, viewing fundamental dimensions of the universe as managerial resources, and embracing paradoxes like mass customizing, are also conceptions that put us on that edge. We must learn to be comfortable out there, for that is where exploration, growth, mastery, and maturation come from. Competitive advantage in the business world lies there, too, and this is where tomorrow's managers must "hang out." From that existential, competitive edge we will be better able to develop transformative ways of managing people in the new economy.

Rationality and creativity, for example, were always thought of as contrasting characteristics, and in large organizations rationality was found more often than creativity. Rational types focus on facts, and facts are about the past. Theory, in contrast, is about the future. Once theory is accepted as fact, it then relates to the past. The more you revere facts, therefore, the more resistant you are to change. The faster things change, the less you can use facts and the more you need imagination.

In banks, portfolio managers can think linearly and sequentially, relying on facts more than the fast-paced foreign exchange traders. Similarly, arbitrage is about 80 percent analytical and 20 percent emotional, whereas trading is the reverse. The best traders are intuitive, relying on feel, imagination, and an inner sense of oneness with currency flows. The thirty-six-year-old head of trading at Bankers Trust, whose bonus is several hundred thousand dollars higher than his CEO's, insists that "right brain people make the best traders. At every moment they see things as a totality." In other words, as machines' mastery of information speeds up, we will need to rely on people less for facts and more for imagination.

The industrial model was geared to bring out the opposite in people. Mass production meant meticulous attention to repetitive detail. Society, therefore, needed to train people out of their natural ability to be playful and creative. Socialization in the industrial world was to make them capable of sustaining boredom in adult life. An example of this is the way children who are naturally creative were punished in schools a few generations back. A not uncommon punishment was to write something like "I will not say ——— again" one hundred times on the blackboard. The purpose of this kind of punishment was to adapt them to carrying out repetitive detail in the industrial labor force. Perhaps today a more creativity-producing punishment for the same act would be to list a hundred other ways saying the same "———" thing.

Recognition of a new economic paradigm will require a new approach to the development of people in organizations. The hierarchical model produced a few creative individuals and a large majority of uncreative people with strong left-brain skills. This was acceptable in an industrial economy where there are a large number of uncreative jobs.

Since many of these jobs will be done by computers in the new economy, the training and development functions in the future will have to shift their emphasis. Education, according to Thomas R. Blakeslee, will have to "concentrate more on development of those skills that are poorly done by computers. Development of creativity and holistic thinking ability should have top priority."[1] It is worth considering what would happen to our businesses and economy if, between now and 2001, all major corporations came to balance rationality and imagination in their training and development programs?

Another "what if" worth considering has to do with people's transformed orientation to their jobs, that is, to the very existence of their jobs. Since downsizing is currently popular, specifically, what would a company look like if there was guaranteed employment, and employees were able to focus on eliminating their jobs, making them obsolete and unnecessary no-matter? Zero-based organization boxes, like zero-based budgeting, justifies all parts and the whole at every moment.

Management, starting with a focus on quality rather than growth, is trying to make more available with less resources, especially people. Almost all downsizing, however, is done hierarchically as well as absolutely. That is, the parts and people to be taken out are almost invariably set by those above them in the authority structure. How many downsizing operations have you ever seen where people focused on eliminating their own jobs, rather than those of others? Not many.

In the chapter Any Time, we spoke of problem solvers who lead meaningful careers whittling away at problems that never go away, partly because of the context in which they hold the "problem." The issue is relevant, here, to the holistic approach we are recommending. The first focus in a holistic downsizing is on the part the person is most familiar with: his or her own job. This would mean a mindset of always looking to reduce and ultimately eliminate your job.

What would a company look like if all employees were focused on eliminating their jobs, making them obsolete and unnecessary? There is a theoretical paradox here, but one that just might work in practical terms. Assume that the company offers guaranteed job security to anyone who can downsize their job to zero. Theoretically, if everyone did this there would be nothing but one big musical chair rotation; truly "full of sound and fury, signifying nothing." Practically, however, it is unrealistic to assume a perfect world where every employee actually buys into the downsizing-with-job-security+

plus guarantee. Those who do, therefore, help themselves and the company, and a percent of those who don't are in the part that is taken out anyway.

The point of this example is only partially about downsizing. Mainly, it is about a transformed mindset among employees: from seeing their responsibility to themselves (keeping their jobs) as being at odds with their job responsibility, to a context in which the part (job) and the whole (company) are the same. The Japanese accomplish this mutual and holistic identification with the essence of their entire culture, in which lifetime employment guarantees (which may be eroding) are only an element. American management is not likely to embrace such notions, and needs to figure out more Western twists in order to forge a new managerial context. And there will be plenty of twists, but the core is now clear: an holistic reality, where universal dimensions become resources applied to science and technology, and offering clues to business, organization, and people for a better life.

We began the book with a progression: universe $\longrightarrow$ science $\longrightarrow$ technology $\longrightarrow$ business $\longrightarrow$ organization. To understand management and organization in the future, I have argued, it is necessary to understand the dimensions of the universe, and how they affect all that follows. This was not a statement of philosophy, and I have worked to show the practical importance of this orientation for management.

In concluding, I would like to come full circle and take one last look at this map for progress. There is a problem that needs resolution before we can end (begin?): universe $\longrightarrow$ science $\longrightarrow$ technology $\longrightarrow$ business $\longrightarrow$ organization is a linear conception with a dead end at "organization." By linear and mechanistic reasoning, the only way for our models of organization to grow and evolve is if we discover the arrow and element that follow "organization." The resolution does not lie in that old standard, the feedback loop, which would be nothing more than a conceptual sleight of hand. In part, the dead end at organization is bureaucracy, and the only way out is by a new beginning with a more encompassing paradigm, as I have tried to outline.

Another resolution is to dissolve the problem through a holistic, rather than a linear, or mechanistic, approach. From a holistic perspective, each part of the progression is the whole progression. Growth and new patterns of organization, therefore, evolve when we grasp the relationship between the parts – as in the way time, space, and mass determine the parameters of science, technology, business, and organization, as well as those of the universe.

The holistic quality of the progression was brought home to me in a meeting with a very unique executive. A bank was started in London, in 1972, with $250 million of Third World oil money. Today, the bank is one of the fastest growing in the world, with assets in excess of $15 billion.

A major U.S. bank once had a 20 percent participation, but the leaders of each institution were from two different worlds; they did not understand what the other was talking about, and the tie was severed. The bank is very global in its orientation and composition, and decidedly non-Western in its organization and culture. No two directors are from the same country, and the president is a very refined Pakistani gentleman in his sixties. He has an aura about him that made me think of Gandhi in banker's pinstripes. Speaking very softly, his almost religious humanism both dominates and obscures a most practical bent.

The president and some of his managers had read parts of this book in draft form, and I was invited to what became a three-hour luncheon dialogue of East-West philosophy applied to business and organization. The setting was elegant, a private bank dining room in the City of London. Every platter was an artistic presentation. The conversation was polite and animated, otherworldly and pragmatic, all at once.

At one point I reviewed the path that I saw from the universe, through science to technology, and then to business, before reaching management and organization. I spoke of the lag that occurs in development from one of these to the next, and of the way in which a linear progression violates notions of holistic management.

Like a teacher to his pupil, he gently whispered: "But Dr. Davis, don't you see that we go *directly* from universe to management!"

I remembered Nobel physicist Paul Dirac's remark, "It is more important to have beauty in one's equations than to have them fit experiment. It seems that if one is working from the point of view of getting beauty into one's equations, and if one has really a sound insight, one is on a sure line of progress." I remembered Commander David Bowman, surviving crew member and Child of the Stars, in *2001: A Space Odyssey*. The progression had been scaffolding, necessary supports along the way toward a new conception for a new economy, and ultimately unnecessary. When management treats time, space, and no-matter as resources rather than as roadblocks, our methods of organization will no longer be lagging behind, at the end.

Notes

ANY TIME
1. Both Weick's and Schutz's quotes are taken from
Karl Weick, *The Social Psychology of Organizing*,
Reading, Mass: Addison-Wesley, 1969.
2. Harold Geneen, *Managing*, New York: Doubleday,
1984.

ANY PLACE
1. Material in this section has been adapted from
"Distribution: A Competitive Weapon," The MAC
Group, Cambridge, Mass., 1985.
2. Marilyn Ferguson, *The Acquarian Conspiracy: Personal and Social Transformation in the 1980s*, Los
Angeles: Tarcher, 1980, p. 213.
3. See Stanley M. Davis and Paul R. Lawrence,
Matrix, Reading Mass: Addison-Wesley, 1977.

NO-MATTER
1. Theodore Levitt, "Marketing Intangible Products
and Product Intangibles," *Harvard Business Review*,
May-June 1981, p. 95.
2. Alfred D. Chandler, Jr., *The Visible Hand*, Cambridge, Mass: Harvard University Press, 1977, p. 1.
3. See Thomas M. Stanback, Jr., et al., *Services: The
New Economy*, Totowa, N.J.: Rowman & Allanheld,
1981, p. 16.

4. See Michael B. Packer, "Measuring the Intangible in Productivity," *Technology Review*, February-March 1983; and Robert S. Kaplan, "Yesterday's Accounting Undermines Production," *Harvard Business Review*, July-August 1984.

5. Arthur Andersen & Co. report to AFSM on "Electronic Products Service Business," AFSM, Ft. Meyers, Fl.

6. James L. Heskett, *Managing in the Service Economy*, Cambridge: Harvard Business School Press, 1986, p. 160.

7. Paul Hawken, *The Next Economy*, New York: Holt, Rinehart & Winston, 1983, p. 11.

8. See G. Lynn Shostack, "How to Design a Service," *European Journal of Marketing*, vol 16, # 1, pp. 49–63.

9. See Theodore Levitt, "After the Sale Is Over ...," *Harvard Business Review*, September-October 1983, pp. 87–93.

10. Alvin Toffler, *The Third Wave*, New York: William Morrow, 1980.

11. For a fuller treatment of this, see the discussion of dissipative structures in Nobel winner Ilya Prigogine's *Order Out of Chaos*, with Isabelle Stengers, NY: Bantam Books, 1984.

12. Daniel Bell, *The Coming of Post-Industrial Society*, New York: Basic Books, 1973.

13. Robert B. Reich, *The Next American Frontier*, New York: Times Books, 1983, p. 141.

14. Karl Albrecht and Ron Zemke, *Service America!*, Homewood, Il: Dow Jones-Irwin, 1985, pp. 6–7.

15. Ibid.

16. See, for example, G. Lynn Shostack, "Designing Services That Deliver," *Harvard Business Review*, January-February 1984.

17. Karl Albrecht and Ron Zemke, *op. cit.*, p. 37.

18. "And Now, The Post-Industrial Corporation," *Business Week* March 3, 1986, p. 64. Italics added.

19. James L. Heskett, *op. cit.*, p. 193. Italics added.

20. "A Productivity Revolution in the Service Sector," *Business Week*, Sept. 5, 1983, p. 106.

21. See Ronald Kent Shelp, *Beyond Industrialization: Ascendancy of the Global Service Economy*, NY: Praeger, 1981.

22. "The Next Trade Crisis May Be Just Around the Corner," *Business Week*, March 19, 1984, p. 48.

MASS CUSTOMIZING

1. Quoted in Marilyn Ferguson, *op. cit.*, pp. 178-179.

2. *SKY*, September 1985, p. 14.

3. "Superfast Computers: You Ain't Seen Nothin' Yet," *Business Week*, August 26, 1985, p. 91, from data gathered by Carnegie-Mellon University and the University of Texas.

4. "Japanese Homebuilders Begin to Export," *Nikkan Kogyo Shimbun*, December 19, 1984.

5. *Wall Street Journal*, April 18, 1985, p. 1.

6. Theodore Levitt, "The Globalization of Markets," *Harvard Business Review*, May-June 1983, pp. 92–102.

7. Ibid., p. 93.

8. Quotes are taken from Gifford Pinchot III, *Intrapreneuring*, NY: Harper & Row, 1985, pp. xi, 232, 252. Italics in original.

9. "How to Succeed at Cloning a Small Business," *Fortune*, October 28, 1985, pp. 60–66.

HELLO FROM THE FUTURE

1. Karl H. Pribram, "'Holism' Could Close Cognition Era," *Monitor*, September 1985, pp. 5–6.

2. Federal Express Corporation, 1984 Annual Report, p. 1.

3. "Redefining an Industry Through Integrated Automation," *Infosystems*, May 19, 1985, p. 27.

4. Quoted in Robert A. Sigafoos, *Absolutely Positively Overnight*, Memphis: St. Luke's Press, 1983, p. 36.

BEFOREMATH

1. Thomas R. Blakeslee, *The Right Brain*, New York: Berkley Books, 1983, p. 113.

Index

236
Index